# THE EAGLE, GLOBE, AND ANCHOR

# 1868-1968

MARINES

HISTORY AND MUSEUMS DIVISION
HEADQUARTERS, U.S. MARINE CORPS
WASHINGTON, D.C.

## MARINE CORPS MUSEUM TECHNICAL MONOGRAPH SERIES

This Technical Monograph Series is prepared under the direction of the Senior Historian and will include other worl<s similar in nature. These are largely derived from the museum,s material collections and are designed to explore in depth the particular subject addressed.

It is hoped that the Series will stimulate interest in the more obscure aspects of the material history of the Marine Corps and at the same time introduce a specialized reference document of considerable utility to the historical researcher as well as the technician.

The Marine Corps Museum is located a few miles south of Washington., D. C., just off Interstate 95 and is open daily to the public free of charge. The research facilities are open to users who have obtained prior authorization from the Curator from 8:00 a. m. to 4:30 p. m. Monday through Friday, federal holidays excepted.

## COVER SEAL

The device reproduced on the front cover is an interpretation of an 1812 United States Marine Corps hat device by JoAnn Wood.

For sale by the Superintendent of Documents, U.S. Government Printing Office
Washington, D.C. 20102

Stock No. 008-055-00117-0

# THE EAGLE, GLOBE, AND ANCHOR 1868-1968

By

Colonel John A. Driscoll, USMCR

PCN 19000319900

HISTORY AND MUSEUMS DIVISION
HEADQUARTERS, U. S. MARINE CORPS
WASHINGTON, D.C.
Printed 1971
Reprinted 1977

For sale by the Superintendent of Documents, U.S. Government Printing om.ce
Washington, D.C. 20402

Stock No, 008-0M--00117-0

# TABLE OF CONTENTS


| CHAPTER | | Page |
|---|---|---|
| | INTRODUCTION | 1 |
| I | 1868-1876 | 5 |
| II | 1876-1892 | 11 |
| III | 1892-1904 | 19 |
| IV | 1904-1914 | 31 |
| V | 1914-1922 | 37 |
| VI | 1922-1930 | 51 |
| VII | 1930-1937 | 61 |
| VIII | 1937-1949 | 67 |
| IX | 1949-1962 | 83 |
| X | 1962-1968 | 97 |

## APPENDICES


| | | Page |
|---|---|---|
| a. | The 1868 Fatigue Cap Ornament. | 103 |
| B. | The 1869 Officer's Epaulette Ornament | 105 |
| C. | The 1875 Uniform Ornaments | 107 |
| D. | The 1892 Uniform Ornaments. | 109 |
| E. | The 1900 Uniform Ornaments. | 111 |
| F. | The 1904 And 1908 Uniform Ornaments. | 113 |
| G. | The 1912 And 1917 Uniform Ornaments. | 117 |
| H. | The 1922 Uniform Ornaments. | 121 |
| I. | The 1929 Uniform Ornaments. | 127 |
| J. | The 1937 Uniform Ornaments. | 135 |
| K. | The 1943 Women's Reserve Uniform Ornaments. | 141 |
| L. | The 1945 Women's Reserve Uniform Ornaments. | 145 |
| M. | The 1949 Uniform Ornaments. | 147 |
| N. | The 1962 Uniform Ornaments. | 157 |
| O. | The 1966 Uniform Ornaments. | 161 |

ILLIISTRA TIONS


IXTRODIT TIOX

| | |
|---|---|
| Figu re 1 | Enlisted hat pla te, circa. 1812 |
| Figu re 2 | Enlisted cap ornament, circa. 1834-1840. |
| Figure 3 | Officer's full dress cap ornament, 1859-1876. |

CH.-\PTER I

| | |
|---|---|
| Figu re 4 | Officer's fatigue cap ornament photograph submitted together with the Boa rds recommendation to Secretary of the Navy Gideon Wells by Brig. Gen. Comdt. Jacob Zeilen on 17 Nov. 1868 for approval. |
| Figu re 5 | Officer's fatigue cap ornament photograph sent to posts and stations on 30 Nov, 1868. |
| Figu re 6 | Officer's fatigue cap ornamen t photograph (rear view) sent to posts and stations on 30 Nov. 1868. |
| Figure 7 | Officer's fatigue cap ornament, 1868-1876. |
| Figu re 8 | Off icer's fatigue cap ornament, 1868-1876. Observe, showing the globe and eagle separated from the dark blue cloth upon which the foul anchor is embroidered in gold bullion, |
| Figure 9 | Officer's fatigue cap ornament, i868-1876. Reverse, showing the globe and eagle separated from the dark blue cloth upon which the foul anchor is embroidered in gold bullion. |

CHA PTER II

| | |
|---|---|
| Figure 10 | "Richly-gilt United States shield" for the officer's full-dress !lat ornament, 1876-1892. |
| Figure 11 | "Silver Corps device" for the officer's full-dress hat ornament, 1876-1892. |
| Figure 12 | Officer in full-dress unifor m presc ribed by the 1875 Uniform Regulations. |
| Figure 13 | Officer's undress cap ornament, 1876-1892 ; officer's fatigue cap ornament, 1876-1881. |
| Figure 14. | Enlisted full-dress hat ornament, 1876-1892. |
| Figure 15 | Enlisted Marine in the full-dress uniform worn from 1876-1891(2). |
| Figure 16 | Enlisted undress cap and fatigue cap ornament illustrated in the 1875 Uniform Regulations. |

ILLUSTRATIONS (Cont'd)


Figure 17 Enlisted Marine in the full-dress coat and undress cap prescribed by the 1875 Uniform Regulations.

Figure 18 Officer's dress coat corps devices, 1877-1892 ; Officer's summer undress coat corps devices, 1892-1900; Officer's undress cu i. corps devices, 1900-1904; Officer's summer undress coat corps devices, 1900-1904; Officer's campaign coat corps devices, 1900-1904.

Figure 19 Musician, U. S. Marine Band, circa. 1885, in full-dress coat and white helmet.

CHAPTER III

Figure 20 Officer in the full-dress uniform worn from 1892-1904.

Figure 21 Officer's undress cap ornament, 1892-1904.

Figure 22 Officer in the undress uniform prescribed by the 1892 Uniform Regulations.

Figure 23 Officer's epaulette ornament (right) 1892-1904.

Figure 24 Officer in the undress coat prescribed by the 1892 Uniform Regulations.

Figure 25 Enlisted black helmet device (type 1), 1892-1904.

Figure 26 Enlisted black helmet device (type 2), 1892-1904.

Figure 27 Enlisted shoulder knots with corps device, 1892-1904.

Figure 28 Enlisted Marine in the full dress coat worn from 1892-1904.

Figure 29 Officer , circa. 1898, in undress uniform.

Figure 30 Officers, 1898, in campaign uniform.

Figure 31 Enlisted Marines, 1900, Peking, China, wearing the undress uniform with campaign hat.

Figure 32 Enlisted Marines, circa. 1900, in the Phillippines, wearing the old blue field kit for tropical use.

Figure 33 Staff officer's shoulder knot ornament (right) , 1900-1904.

Figure 34 Enlisted cap ornament, 1900-1908.

CHAPTER IV

Figure 35 Officer's service cap and field hat ornament, 1904-1914.

ILLUSTRATIONS (Cont'd)


| | |
|---|---|
| Figure 36 | Officers, September 1906, Camp Columbia, Mariano, Havana Province, Cuba, shown in field uniform of the period. |
| Figure 37 | Officer, circa. 1909, in blue undress uniform. |
| Figure 38 | Officer's service collar ornament (right), 1904-1914. |
| Figure 39 | Officer, 1910, in the field coat and khaki cap of the period. |
| Figure 40 | Enlisted Marines, circa. 1910, shown in the undress uniform worn from 1904 to 1914. |
| Figure 41 | Enlisted Marine, circa. 1910, in the prescribed field uniform of the period. |

CHAPTER V

| | |
|---|---|
| Figures 42 & 43 | Officers full-dress and undress cap ornament, 1914-1919; officers dress cap ornament, 1919-1937. |
| Figure 44 | Commandant's and Estaff officer's chapeau ornament, 1914-1919. |
| Figure 45 | Officer's field cap and hat ornament, 1914-1922; officers bronze cap and hat ornament, 1922-1930; officers service cap and hat ornament, 1930-1937. |
| Figure 46 | Officer, 1912, in winter field uniform. |
| Figures 47 & 48 | Officer's undress collar ornaments, 1914-1919; Officer's dress collar ornaments, 1919-1922; Officer's dress collar, epaulet and shoulder knot ornaments, 1922-1942; Officer's dress collar and shoulder knot ornaments, 1942-1948; Officer's dress collar ornaments, 1948-1962. |
| Figure 49 | Officer, 1917, in white undress uniform. |
| Figure 50 | Officer's field coat collar ornaments, 1914-1922; Officer's bronze collar ornaments, 1922-1927. |
| Figure 51 | Officer's epaulet ornaments, 1914-1919. |
| Figure 52 | Officer, 1916, in full-dress coat with epaulets. |
| Figures 53 & 54 | Enlisted dress cap ornament, 1914-1922. |
| Figure 55 | Marine Gunnery Sergeant, circa. 1919, in undress uniform. |
| Figures 56 & 57 | Enlisted field cap and hat ornament, 1914-1922. |

Figure 58 Marine Sergeant, circa 1919, in winter field coat and field hat.

Figure 59 Enlisted service collar ornaments, 1918-1919.

Figure 60. Officer , circa. 1918, France, in winter field uniform with overseas cap.

Figure 61 Enlisted Woman Reservist, circa. 1918, in winter field uniform with field hat.

Figure 62 Enlisted Woman Reservist, circa. 1918, in winter field uniform with winter field Glengarry cap.

Figure 63 Enlisted dress coat collar ornaments, 1920-1930.

Figure 64 Enlisted service coat collar ornaments, 1920-1930.

CHAPTER VI

Figure 65 Marine Corporal, circa. 1922 , in dress uniform.

Figure 66 General Officer , 1922, in evening dress uniform.

Figure 67 Commandant, circa. 1922-1924, in dress uniform.

Figures 68 & 69 Enlisted dress cap ornament, 1922-1930.

Figure 70. Enlisted bronze cap and hat ornament, 1922-1930.

Figure 71 Enlisted Marine, circa. 1922 , in winter service uniform with field hat.

Figure 72 Second Leader , United States Marine Band, 1922 , in full dress uniform.

Figure 73 Enlisted Marines, 1928, Tientsin, China, in winter service uniform with roll-collar coat.

Figure 74 Illustration accompanying Circular Letter No. 46 of December 1926, Roll-collar coats, Officers. Manner of placing ornaments on collar and insignia on lapel.

Figure 75 Standard U. S. Marine Corps Emblem, approved 28 May 1925.

CHAPTER VII

Figure 76 Officer , 1936 , in blue undress B uniform.

Figure 77 Officer's service collar ornament (left), 1927-1937.

ILLUSTRATIONS (Cont'd)


| | |
|---|---|
| Figure 78 | Officer , circa. 1936, in winter service C uniform. |
| Figure 79 | Enlisted gilt cap ornament, 1930-1937. |
| Figure 80 | Enlisted bronze cap and hat ornament, 1930-1937. |
| Figure 81 | Enlisted gilt collar ornaments, 1930-1937. |
| Figure 82 | Enlisted bronze collar ornament (right) , 1930-1937. |
| CHAPTER VIII | |
| Figure 84 | Standard U. S. Marine Corps Emblem, approved 16 March 1936. |
| Figures 85, 86 & 87 | Officer's dress cap ornament, 1937-1962. |
| Figures 88, 89 & 90 | Officer's service cap and hat ornament, 1937-1942; Officer's service cap ornament, 1942-1962. |
| Figures 91 & 92 | Officer's dress collar ornaments, 1937-1962. |
| Figure 93 | Leader , Marine Band, 1937, in full dress uniform. |
| Figures 94 & 95 | Officer's service collar ornaments, 1937-1962. |
| Figures 96 | Enlisted gilt cap ornament, 1937-1949; enlisted dress cap ornament, 1949-1955. |
| Figure 97 | Marine Corporal, circa. 1945, in blue undress "B" uniform. |
| Figure 98 | Enlisted bronze cap and hat ornament, 1937-1942; Enlisted bronze cap ornament, 1942-1955. |
| Figure 99 | Enlisted gilt collar ornaments (screw post type), 1937-1955. |
| Figure 100 | Enlisted bronze collar ornaments (screw post type) , 1937-1949; Enlisted service collar ornaments (screw post type), 1949-1955. |
| Figure 101 | Marine Pfc, circa. 1945, in winter service "A" uniform. |
| Figure 102 | Mar ine Corporal , circa. 1945, in summer service unif orm with fiber helmet. |
| Figure 103 | Enlisted plastic service cap ornament, circa. 1943 |
| Figure 104 | Enlisted plastic service collar ornaments, circa. 1943. |
| Figure 105 | Service collar ornam-ent (right) (obverse) , circa. 1943, procured in Australia. |

ILLUSTRATIONS (Cont'd)


Figure 106 Service collar ornament (right)( reverse) , circa. 1943, procured in Australia. Marked "K. O. LUKE MELBOURNE".

Figure 107 Marine First Sergeant, 1946 , in new winter service "A" uniform with green wool service jacket and winter service garrison cap.

Figure 108 Marine Pfc., 1948, in summer service A uniform.

Figure 109 Marine First Sergeant, 1946, in blue undress "A" uniform.

Figure 110 Woman Reserve Officer, 1944, in summer service "A" uniform.

Figure 111 Woman Reserve Corporal, 1943, in winter service uniform.

Figure 112 Woman Reserve Sergeant, 1944, in summer dress uniform.

Figure 113 Woman Reserve Officer, 1944, in summer dress A uniform.


**CHAPTER IX**


Figure 114 Enlisted service collar ornaments (clutch) , (obverse) , 1951-1955.

Figure 115 Enlisted service collar ornaments (clutch) , (reverse) , 1951-1955.

Figure 116 Enlisted gilt collar ornaments (clutch) , circa. 1951-1955.

Figure 117 Woman Marine Officer , 1952 , in blue undress uniform.

Figure 118 Woman Marine Captain, 1952 , in winter service "A" uniform.

Figure 119 Enlisted Women Marines uniforms, 1952. Blue undress uniform, winter service "A" uniform and summer service uniform.

Figure 120 Correct placement of enlisted collar ornaments on shirt, khaki, with quarter length sleeve.

Figure 121 Seal of the United States Marine Corps, approved 22 June 1954 by President Dwight D. Eisenhower.

Figure 122 Marine Corporal, 1960, in summer service "A" uniform.

Figure 123 Enlisted dress cap ornament, 1955-1956; enlisted dress cap insignia, branch of service, 1956 to the present.

Figure 124 Enlisted bronze cap ornament, 1955-1956; enlisted bronze cap insignia, branch of service, 1956-1961; enlisted service cap insignia, branch of service, 1961-1963.

Figure 125 Enlisted gilt collar ornaments (clutch) , 1955-1956; enlisted gold collar insignia, branch of service (clutch) , 1956-1961; enlisted dress collar insignia, branch of service (clutch) , 1961-1961.

ILLUSTRATIONS (Cont'd)

Figure 126 Enlisted gilt collar ornaments (screw post) 1955-1956; enlisted gold collar insignia, branch of service (screw post) , 1956-1961; enlisted dress collar insignia, branch of service, 1961 to the present.

Figure 127 Enlisted service collar ornaments (clutch) , 1955-1956; enlisted bronze collar insignia, branch of service (clutch) , 1956-1961; enlisted service collar insignia, branch of service (clutch) , 1961-1963.

Figure 128 Officer's dress cap insignia, branch of service, 1962 to the present.

Figure 129 Officer's service cap insignia, branch of service, 1962-1963.

**Figure** 130 Officer's dress collar insignia, branch of service 1962 to the present.

Figure 131 Officer's bronze collar insignia, branch of service, 1962-1963.

CHAPTER X

Figure 132 Musician, U. S. Marine Band, 1965, in full dress uniform.

Figure 133 Officer's service cap insignia, branch of service, 1963 to the present.

Figure 134 Officer's service collar insignia, branch of service, 1963 to the present.

Figure 135 Enlisted service cap insignia, branch of service 1963 to the present.

Figure 136 Enlisted service collar insignia, branch of service, 1963 to the present.

# INTRODUCTION

.For the past century the E agle, Globe, and Anchor emblem has been the symbol of the United States Marines. The purpose of this study is to explore and describe the development of the emblem and to provide students of Marine Corps history with a reference for its display on the diversity ofuniforms worn by Marines since 1868.

Evidence of emblems or ornaments ascribed to the Marine Corps can be found in correspondence as early as 1804 which refers to a br ass eagle on a square plate being worn, while an octagonal br ass plate embossed with an eagle, foul anchor , drums, flags, etc. was worn during the period of the War of 1812. (Fig. 1)

Subsequent early devices include a gold plate with a gold eagle for wear with the of ficer's full-dress of 1821, while on 4 November 1834 an order prescribed a br ass eagle "measuring from three to three and one-half inches from one ex- tremity of the wing to the other", for wear on caps. (Fig. 2)

During the early part of 1840 , the undress hats of both officers and enlisted men displayed gold wreaths encircling the letters "U.S.M." A f ew months later , a blue cloth cap with black patent leather visor and "with a foul anchor of gold, in a gold wreath in front" was prescribed for dficers not in full-dress uniform.

In 1859 another ornament was adopted - a yellow metal huntinghorn, within which was placed a white metal letter "M". For the full-dress cap, the bugle with "M" was mounted on a U.S. shield which was worn together with a laurel wreath. Both shield and wreath were of yellow metal. (Fig. 3)

The ornament for the undress cap consisted of the yellow metal (or gold embroidered) bugle with the white metal (or silver) letter "M" while the ornament worn on the Officer's epaulette consisted of a solid silver bugle with a solid silver or plated "M".

As the bugle of the 1859 ornament closely resembled the insignia of the U. S. Army inf antry of that perioa, a step was taken in 1867 to provide the Marine Corps with a truly distinctive device. An order dated 2 May of that year directed that an ornament consisting of a "silver Western Hemisphere with gold continents and sur- mounted by a silver eagle" be worn on the officer's undress cap. This order was short-lived, however , as it was suddenly revoked on 5 June 1867.

Finally, iil November 1868, an "Eagle, Globe and Anchor" ornament was pre- scribed, which in spite of various changes in design, size and color , has survived the succeeding century unchanged in concept

Although over the past 100 years this emblem has been displayed on such items as aircraf t, belt plates, flags, guidons, helmet covers, official correspondence and publications, saddle cloths, tie clasps and utility clothing, this study deals only with the metal uniform ornament, its changes, and the article of uniform clothing on which it has been worn.

As in any study of this type, dependent of necessity on fragmented records, it is recognized that more remains to be learned as additional information comes to light.

Corrections and additions undoubtedly will have to be made in the future to supplement this monograph. It is hoped, however , that this study will serve as a useful and convenient reference on its 3ubject, as well as being a vehicle for bringing previously unpublicized facts fore.

Since many of the users of this type of publication find the actual text of official documents particularly useful, an extensive appendix has been included. These reproduce the pertinent sections of a number of uniform regulations. Also, for reference purposes, the illustrations, where practical, have been reproduced in actual scale so that the reader may compare them with origin al specimens for identification. In addition , a brief illustrated glossary is provided to simplify understanding the frequently technical terminology used in a study of this nature. It should be noted that the spellings used are those of the particular time under discussion and thus epaulette and epaulet as well as other divergent spellings are occasionally used to maintain continuity with the documents being quoted or paraphrased.

Figure 1. - Enlisted Hat Plate, circa 1812.

Figure 2. - Enlisted Cap Ornament, circa 1834-1840.

Figure 3. - Officers Full-Dress Cap Ornament, 1859-1876.

# CHAPTER I

On 12 November 1868, Brigadier General Commandant Jacob Zeilin, then the senior officer of the U. S. Marines, appointed a board consisting of Major George R. Graham, Captain Robert W. Huntington and Brevet Captain Charles F. Williams and directed them as follows:

> You are hereby constituted a Board for the purpose of deciding, and re- the porting upon various devices of Cap ornaments proposed for the Marine Corps.
>
> You will assemble on Friday morning, the 13th Inst., and report to me in writing which of the devices submitted is best suited for adaptation in the Corps.

The three member board convened as directed and subsequently submitted the following report to the Commandant:

> In obedience to your order of the 12th Inst., hereto annexed, we have examined the several patterns of devices presented as a substitute for the ornament now worn on the fatigue cap of the Officers and enlisted men of the United States Marine Corps and have respectfully recommended the following as the most appropriate for adoption:
>
> For Commissioned Officers: - A frosted silver hemisphere struck from solid plate with chased parallels and continent of North and South America of gold plate; to be surmounted by a spread eagle cut from solid silver and securely fastened by means of a tang of silver soldered on the inner side of the hemisphere. The hemisphere to rest upon the shank, between the stock and flukes, of a foul anchor worked in gold bullion. The dimensions to be as follows: - hemisphere three fourths (3/4) of an inch in diameter , the eagle three eighths (3/8) of an inch from top of head to point of claws, spread of wings one (1) inch. Shank of anchor to intersect the hemisphere between the Arctic Circle and the Tropic of Cancer on the left and the Equator and Tropic of Capricorn on the right. Anchor over all in length one and one-half (1 1/2) inches and one eighth (1/8) of an inch in width, stock five eighths (5/8) of an inch in length, one eighth (1/8) of an inch in width, the inner edge of flukes and stock touching the hemisphere; the arms of anchor including flukes to be eleven sixteenths (11/16) of an inch in length from point of fluke to point of shank and one eighth (1/8) of an inch in width. The whole to be on a ground work of dark blue cloth. We enclose herewith a drawing of accurate dimensions of the design recommended.
>
> For enlisted men, same as for Officers except that the whole be struck from plate brass 1/16 inch thick.
>
> (The phrase "1/16 thick" was added to the original letter, in a script differing from that of the body of the letter.)

The Commandant concurred with the board's recommendations and on 17 November sent them on to Secretary of the Navy Gideon Welles, for approval:

I respectfully submit to the Department the enclosed proceeei" 6s of a Board of Officers, convened at this station in pursuance of the annexed order.

I also transmit a colored photogr aph likeness of the device recommended by the Board for adoption.

I approve of the proceedings, and would respectf ully ask the authority of. the Department for the adoption, in the Corps, of the C ap ornament recommended by the Board. (Fig. 4)

Two days later, Secretary Welles informed General Zeilin that "The Department authorizes the adoption, in the Marine Corps, of the cap ornament recommended by the Board of Officers, whose repor t accompanied your letter of the 17th Inst."

With the approval of this ornament, the Adjutant and Inspector's Office on 30 November notified the Commanding Officers of eleven Marine Corps posts and stations throughout the United States that:

> The Fatigue Cap Ornament of the M arine Corps having been changed by order of the Secretary of the N avy, I enclose herewith a copy of the descrip- tion of the one adopted, together with a photograph of the same, in order that you may have it adopted by your Command as soon as practicable" • • • •

"Fig. 1. Photograph of Cap Ornament, full size."

"Fig. 2. Photograph of Section of Ornament, "a" shows manner of fastening the Eagle to the globe, "b" shows tang for securing the globe to the cap, or cloth on which the anchor is worked." (Figs. 5 & 6)

The actual use of the new ornament by enlisted men was deferred until the new fiscal year beginning on 1 July 1869. A contract effective that date with Bent and Bush, a military accoutrement manufacturer and supplier of Boston, Mass. to "supply and furnish to the Marine Corps• • • • 3,000 Fatigue Cap Ornament (at a price of) fif ty cents each.

Although a source of supply for officer ornaments, had not been established, undoubtedly officers had their ornaments made up by local jewelers since photographs were included with the order of 30 November 1868.

Until recently no remaining specimens of the 1868 ornament were known to exist, but in 1967, an off icers fatigue cap emblem was discovered at the Marine Corps Museum, Quantico, Virginia among the recently acquired memorabilia of Brigadier General Henry Clay Cochr ane, USMC, (1842-1913). (Figures 7, 8, and 9)

The ornament adopted in 1868 was still a step or two away from being the standard emblem worn by M arines. The ornament as originally adopted was for wear on the fatigue cap, while the bugle with "M" ornament was still to be worn on both the full-dress cap and officer's epaulettes.

As a step toward standardization the new ornament replaced the bugle with "M" on the officer's epaulette on 20 November 1869 when the following order was published by the Adjutant and Inspector's Office:

By direction of the Navy Department, there will be substituted for the Silver Bugle worn by the officers of the Marine Corps on the Epaulette, an ornament of the same design as that at present worn by Officers on the fatigue cap, except that the anchor be of plain silver, and the ornament to be of the following dimensions:

| | |
|---|---|
| Length of anchor, over all . . . . . . . . . . | 2 inches |
| Length of anchor stock . . . . . . . . . . . . . | 1/2 inch |
| Diameter of globe . . . . . . . . . . . . . . . | 9/10ths of an inch |
| Height of eagle . . . . . . . . . . . . . . . . | 4/10ths of an inch |
| Spread of wings, from tip to tip . . . . . . . . . . | 1 inch |

Deviations from regulations having been observed in the cap ornament, as also in the size, shape, and style of gold cord of the shoulder knots worn by many officers, attention is called to the importance of uniformity; and Commanding Officers will require strict compliance with the patterns and designs prescribed for the Corps.

Finally, with the issuing of new Uniform Regulations in May 1875 (with an effective date of 1 July 1876), the Eagle, Globe, and Anchor ornament became the sole emblem of the United States Marine Corps. At this writing, the design of the Marine Emblem cannot be credited to any single individual, but the 1868 Board evidently considered ornament designs submitted by several people. Second Lieutenant George C. Reid, Aide-de-Camp to the Commandant, is thought to have submitted one and it is likely that at least Lieutenants C. L. Daniels and George M. Welles did so as well. There is as yet, no positive evidence as to who conceived the successful design and so for the moment the problem remains unresolved.

Figure 4. - Officers Fatigue Cap Ornament.
Photogr aph submitted to the Secretary of the Navy on 17 November 1868.

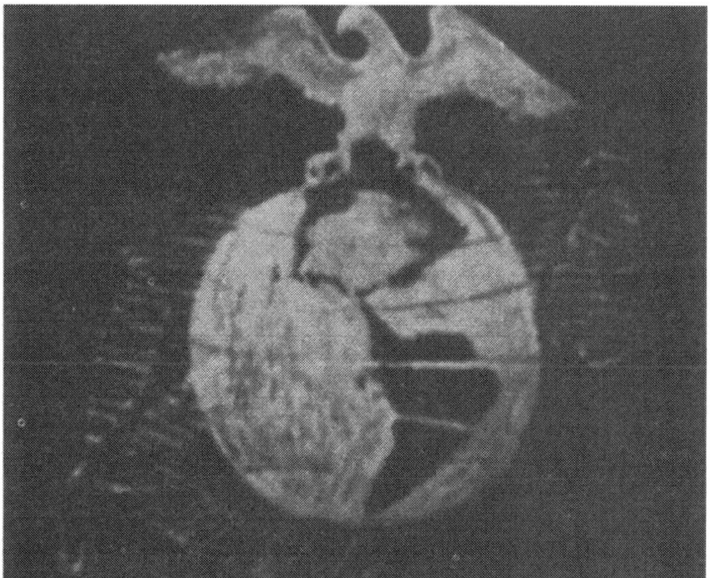

Figure 5. - Officer's Fatigue Cap Ornament.
Photograph sent to posts md stations on 30 November 1868.

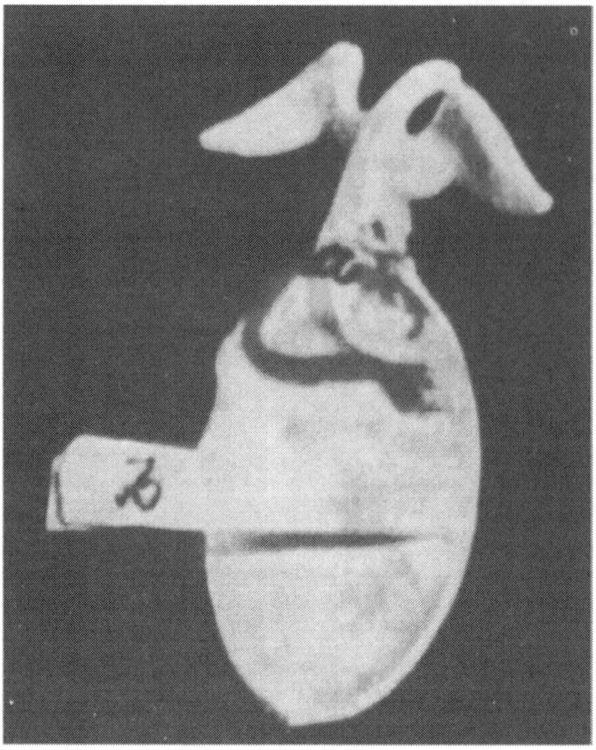

Figure 6, - Officer's Fatigue Cap Ornament.
Photograph (rear view) sent to posts and stations on 30 November 1868,

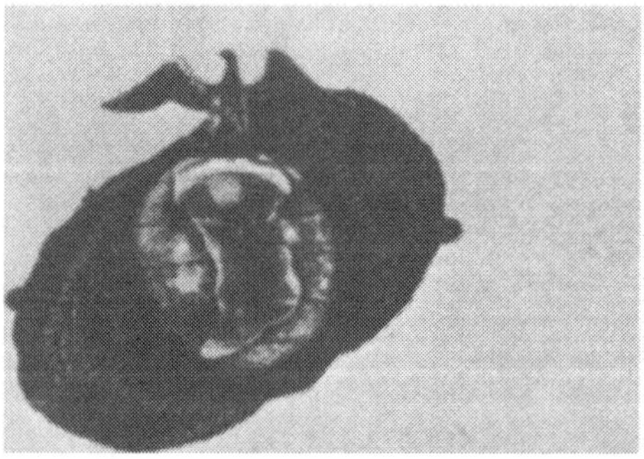

Figure 7, - Officer's Fatigue Cap Ornament, 1868-1876,

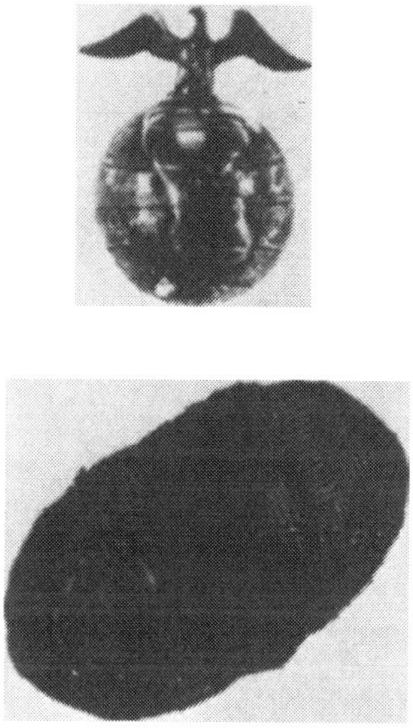

Figure 8. - Officer's Fatigue Cap Ornament, 1868-1876. (Front)

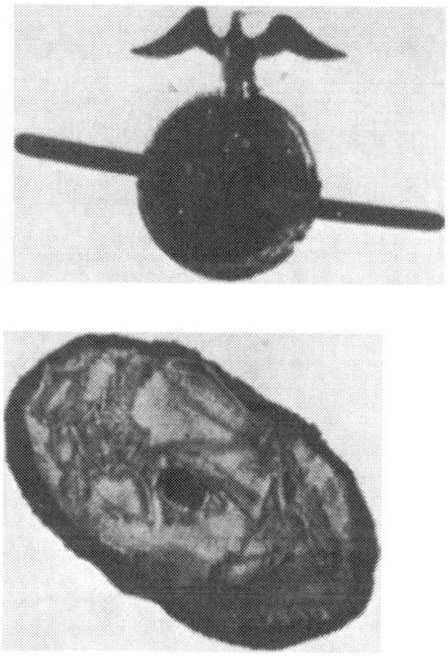

Figure 9. -Officer's Fatigue Cap Ornament, 1868-1876. (Back)

## CHAPTER II

On 16 December 1872, Lieutenants Henry Clay Cochrane and Richard Renshaw Neill of the U. S. Marine Corps, circulated to every accessible officer of the Corps, a neatly printed unofficial letter, requesting each officer's views as to the changes and modifications he desired in the uniform that had then been in use for fourteen years. The letter was well received for at least fif ty replies were returned by the middle of the following February.

In a letter to the Editor of the ARMY AND NAVY JOURNAL published in the 22 February 1873 issue, Lieutenant Cochrane proposed a complete new set of Uniform Regulations largely derived froin the results of his survey.

Regarding the ornament to be worn on the officer's full-dress cap, Cochrane suggested that it should be:

> • • • • a gilt ornament halhhe size of the present one, embodying the present corps cap device within a deeply cut laurel and oak wreath, and if desirable, with a scroll at bottom for an inscription, say "USMC - 1798" (the date of establishment) • • • •

The ornament for the officers "epaulet" was to be "as now worn• •••" and the fatugue cap for both officers and enlisted men was to display the revised emblem.

Eventually Lieutenant Cochrane's efforts bore fruit for on 4 April 1874 the ARMY AND NAVY JOURNAL printed an article which stated in part:

> Upon the written application of Lieutenant Henry C. Cochrane, representing the almost unanimous wish of the field, staff arid company officers of the Marine Corps, General Zeilin, the Commandant, has decided to convene a board for the purpose of revising and improving the uniform of the corps, which has been in use, with slight modification, since 1859.

The board convened as directed and subsequently proposed new Uniform Regulations that were approved by Secretary of the Navy Robeson on 27 May 1875 "to go into effect July 1, 1876."

The new regulations prescribed that the officer's full-dress hat ornament should be:

> "a richly-gilt United States shield, two and seven-eighths inches long by two and three-eighths inches wide; resting upon the centre of the shield a silver corps device of same design and dimensions as that prescribed for the officers undress cap. (Figs. 10, 11 & 12)

The ornament for the undress cap for field and company officers and for the company officer's fatigue cap was described as:

> A fretted silver western Hemisphere, struck from solid plate, with chased parallels; continents of North and South America of gold plate. To be sur- mounted by a spread eagle cut from solid silver , and securely fastened by a tag

silver soldered on the inner side of the hemisphere. The hemisphere to rest upon the shank, between the stock and flukes of a foul anchor embroidered in gold bullion; the inner edges of the flukes touching the hemisphere. The dimensions to be as follows: hemisphere, three-fourths of an inch in diameter; the eagle, three- eighths of an inch from top of head to point of claws; spread of wings, one inch; width of anchor stock and shank, one-eighth of an inch; arms of anchor including flukes, three-fourths of an inch in length from the peak of flukes to apex of crown, and one-eighth of an inch in width; length of anchor over all, one and one-half inches. The shank of the anchor to intersect the hemisphere between the Arctic Circle and the Tropic of Cancer on the left, and Equator and Tropic of Capricorn on the right. The whole to be on a ground of dark blue cloth. (Fig. 13)

The 1875 Uniform Regulations also prescribed an ornament for the officers' epaulette to be a: "Device of the corps in silver, placed within the crescent".

In addition ornaments for the enlisted full-dress hat and the undress and fatigue caps were also prescribed. For the full-dress hat:

"a brass United States shield, two and seven-eighths inches long by two and three-eighths inches wide; resting upon the centre of the shield a nickle-plated corps device, cut from one piece of brass, of the same design and dimensions as prescribed for the officers full-dress hat." (Figs. 14 & 15)

The ornament worn on both the enlisted undress cap and the fatigue cap was described as:

"the same design and dimensions as that prescribed for officers, but struck from one piece of brass and fastened to the front of the cap by two brasswire loops soldered on the inner side of the anchor." (Figs. 16 & 17)

In 1877 by virtue of General Order No. 4, collar ornaments were instituted as part of the officer's uniform for the first time, and since have been worn continuously. The new order which was dated 29 October authorized an evening dress uniform which included a coat of dark blue cloth, displaying a Corps device "one half the size prescribed for the epaulette" on the ends of the collar. (Fig. 18)

It should be noted that these collar devices did not contain an anchor rope or cable. The enlisted collar devices were to obtain the rope in 1955 while that for the officer's device did not appear until 1962.

General Order No. 2 of 10 March 1881 described a new white helmet to be worn by officers, and a handwritten directive bearing the Quartermaster's Office date stamp of 17 March 1881, contained specifications or an enlisted white helmet to replace the fatigue cap with white covers which had been discarded the previous September. The descriptions for the officers and enlisted white helmets do not specify ornaments but contemporary photographs show the U. S. Shield and Corps device for the full-dress hat being worn on the helmet. (Figs. 10, 11, 14 & 19)

In 1891, the shako passed from the Marine Corps and was replaced by a black helmet of the Prussian style, which became popular in military wear after Prussia's tremendous successes against Austria and France, 1866-1871. At first, the shield and Corps device from the shako was specified for use on the new helmet, but shortly, a large ornament incorporating the eagle, globe, and anchor and deleting the shield was designed to supercede the older item. Details of the revised ornament are included in the following chapter.

Figure 10. -"Richly-gilt United States shield" for the Officer's full-dress hat ornament, 1876-1892 .

Figure 11. -"Silver Corps device" for the officer's full-dress hat ornament. 1876-1892.

Figure 12. - First Lieutenant Henry H. Costin is seen in the full-dress uniform prescribed by the 1875 Uniform Regulations.

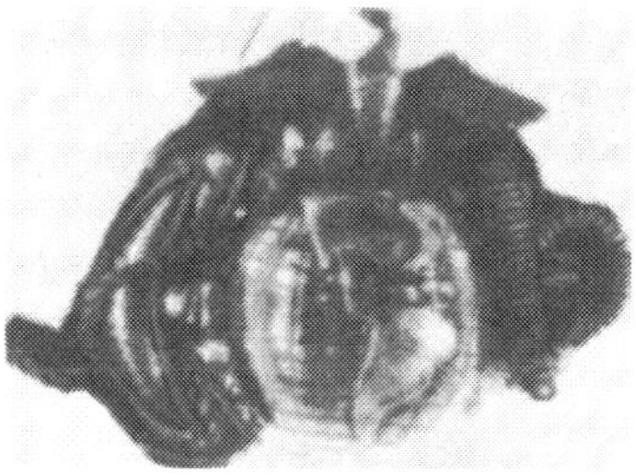

Figure 13. - O f f  icer's Undress Cap Ornament, 1876-1892, and Officer's Fatigue Cap Ornament, 1876-1881.

Figure 14. - Enlisted full-dress hat ornament, 1876-1892.

Figure 15. - Enhsted Marine in the full-dress uniform worn from 1876 to 1891 (2)

Figure 16. - Enlisted undress cap ornament 1876-1892 ; enlisted fatigue cap ornament, 1876-1881 as illustrated in the 1875 Uniform Regulations.

Figure 17. - Enlisted Marine private in the full-dress coat and undress cap, 1875 Uniform Regulations.

Figure 18. - Officer's dress coat corps devices, 1877-1892 ; Officer's swnmer undress coat corps devices, 1892-1900; Officer's undress coat corps devices; 1900-1904; Officer's swnmer undress coat corps devices, 1900-1904; Officer's campaign coat corps devices, 1900-1904.

Figure 19. - Musician, U, S. Marine Band,
circa. 188&. in full dress coat and white helmet with "ornament on front of hat."

# CHAPTER III

Acting on the recommendations of a Unifor m Board that convened in 1892, Colonel Commandant Charles Heywood sent a newly draf ted set of Unif orm Regula- tions on the Secretary of the Navy, B. F. Tracy for approval. These were soon endorsed and resulted in General Order No. 3 and the Commandant's covering letter to the new regulations, both of which emanated from Headquarters U. S. Marine Corps, and bore the date of 14 July 1892.

General Order No. 3 set forth the changes in regulations for officers uniforms only, while the Commandant's letter related to new regulations for both of ficers and enlisted men. Both the Gener al Order and the covering letter stated that the new regulations would be eff ective 15 September 1892.

Following the lead of their predecessors, the new regulations set forth the de- scription of ornaments for both officers and enlisted uniforms. Officers ornaments were prescribed for the black helmet, the undress and white caps, the epaulettes and the undress 0oat, while ornaments for enlisted men were for the black helmet, the undress cap and the shoulder knots. An undress coat collar ornament, also a Corps device, was specified exclusively for the enlisted members of the Band.

The ornament for the officer's black helmet was described as "the same pattern as adopted for enlisted men, with fretted silver Western Hemisphere, with chased parallels; continent of North and South America, and anchor to be of gold, eagle to be of silver."

The of ficer's helmet device ( Fig 20) was, except f or its silver and gold color combination, identical to the enlisted helmet devices shown in Figs. 25 & 26.

The officer's undress and white cap ornament was:

> "A fretted silver western hemisphere, struck from solid plates with chased parallels; continents of North and South America of gold plate. To be surmounted by a spread eagle cut from solid silver , and securely fastened by a tag of silver soldered on the inner side of the hemisphere. The hemisphere to rest upon the shank, between the stock and flukes of a foul anchor in gold metal; the inner edges of the flukes nearly touching the hemisphere. The dimensions to be as follows: hemisphere, three-fourths of an inch in diameter; the eagle, three- eights of an inch from top of head to point of claws; spread of wings, one inch; width of anchor stock and shank, one-eighth of an inch, arms of anchor , including flukes, three-fourths of an inch in length from peak of flukes to apex of crown, and one-eighth of an inch in width; length of anchor over all, one and one-half inches. The shank of the anchor to intersect the hemisphere between the Artie Circle and the Tropic of Cancer on the lef t, and Equator and Tropic of Capri- corn on the right." (Figs. 21 & 22)

The ornament for the Commandant's undress cap only was the "Corps device set in a wreath of gold acorn leaves", "while the officer's epaulette ornament was a "Corps device, in silver placed with the crescent" of the epaulettes. (Fig. 23)

For the undress coat of officers the ornament was to be:

"of metal and one half the size adopted for epaulettes • • • to be worn on each side of the collar ."

This Corps device is the same as the ornament shown in Fig. 18 of the preceeding chapter. (Fig. 24)

With the exception of its metallic content, the enlisted black helmet ornament was identical to that prescribed for officers. The ornament was:

"An enlarged Corps device, consisting of the eagle, globe, and anchor ; to be made of brass, according to the standard device in the Quartermaster's office." (Figures 25 & 26)

The enlisted undress cap ornament was described as being "of the same pattern and dimensions as the present standard sample." See Fig. 17 of the previous chapter.

With the promulgation of the 1892 Regulations, shoulder knots with ornaments were first prescribed for enlisted men and were worn until 1904. These displayed "nickel plated Corps devices" worn in the center of the shoulder knot pads. (Fig. 27 & 28)

In 1892 Uniform Regulations also provided for collar ornaments for the undress coats of the enlisted members of the band. The collar ornament for the undress coat of the Second leader of the band was "a small Corps device of white metal" that was worn "near each point of the collar." The musician undress coats were to have "near each (collar) point , a small Corps device of white metal," similar to that worn by officers of this period. ( Figs. 18 & 24) These devices which were replaced by small lyres of white metal or gold by the 1900 Regulations, represented the first instance devices were worn on the collar by enlisted marines.

In 1897 the chasseur ( French Kepi) pattern undress cap of the Civil War era was discarded and was replaced by a cap of a type the U. S. Army had previously adopted, however , the device worn on the new headgear remained the same as that which it superceeded. (Fig. 29)

On 22 April 1898, the day after war was declared with Spain, the First Battalion of Marines, commanded by Lieutenant Colonel Robert W. Huntington, embarked from Brooklyn, N. Y. on the USS PANTHER for a staging area at Key West, Florida. After additional training at Key West, the battalion again boarded the PANTHER , from which it made its famous landing at Guantanamo, Cuba on 10 June 1898.

Colonel Commandant Heywood in his report of 24 September 1898 to the Secretary of the Navy stated that: "The greatest care was exercised in fitting out the battalion . . . . campaign suits of brown linen and campaign hats were ordered, but owing to the great demand for these articles at the time by the Army, it was impossible to send them with the battalion. They were shipped later , however , and proved a great comfort to the men."

The brown linen campaign suits were the forerunner of the soon to be issued khaki campaign uniforms, while the wide brimmed drab felt campaign hat with a "fore and aft" crease, was similar to that which was adopted by the U. S. Army in 1885. Contemporary photographs indicate that both officers and enlisted Marines wore their undress cap ornaments on the campaign hat and while a very few of them show the ornament being worn on the front of the hat, the majority indicate that the ornament was more generally worn on the left side in the center of the crown. (Figs. 30 & 31) Later the ornament was shifted to the front center and worn there exclusively.

In 1900 a new set of Uniform Regulations was published which contained all of the 1892 Regulations with changes, as well as a few additional ones that were not previously covered, but the specifications for all officers and enlisted ornaments were almost identical to those of the 1892 Regulations. Some small changes, however, were made including the substitution of the Corps device for the small metallic "Coat of Arms of the United States" Army insignia worn on the Chapeau of the Commandant and Staff officers. The new Corps device was "a gilt and burnished Marine Corps Cap ornament" which was placed on a 2 1/4 inch gold lace strip on the right side of the chapeau.

Another change brought about by these regulations provided for Corps device to be worn on the collar of the officers summer undress coat "in the same manner prescribed for (the) blue cloth undress coat." In addition, the "Corps device prescribed for the summer undress coat" were worn on the collar of the officers campaign coat. These ornaments are the same that were prescribed in 1877 and were to be worn until 1904. (Fig. 18)

These new regulations were devoid of comment, probably through oversight, regarding a description of the officers campaign hat and its ornament, as well as information as to how the ornament was to be positioned on the hat. However , it is evident from contemporary photographs that officers wore a hat similar to the one worn by enlisted men and the undress cap ornament was worn positioned in the front of the hat. (Fig. 32)

All regulations from 1869 through 1900 dealing with officers' epaulettes and shoulder knots, prescribed both rank insignia and Corps device for the epaulettes, while the rank insignia only was to be worn on the shoulder knots. Since Staff Officers did not wear epaulettes but used shoulder knots instead, the Corps device was required for staff shoulder knots. The 1900 ornament for the Staff Officer shoulder knots was: "• • • • a corps device of gold and silver, of size midway between the collar and cap ornaments, to be worn on end of knot, flukes of anchor pointing to the front." (Fig. 33)

The 1900 Uniform Regulations also brought about a change in the description of the enlisted cap ornament. Previously the 1892 Regulations described the undress cap ornament as being "of the same pattern and dimensions as the present standard sample." The new regulations directed that the cap ornament should be: "A Corps device of gilders' metal No. 14 , of the same pattern and dimensions as the present standard sample in the Quartermaster's Department."

This ornament was worn on the undress cap and the campaign hat. (Fig. 34)

Figure 20. - Captain Henry Clay Cochrane, 1895, shown in the full-dress uniform worn from 1892-1904.

Figure 21. - Officer's undress Cap ornament, 1892-1904.

Figure . - . L U a..., u,..., ••• • ...,, .L U v"% . J.U a..L HiC: Dcl..L".l"aCKS, Boston, Mass., shown in the undress uniform presc ribed by the 1892 Unifor m Regulations.

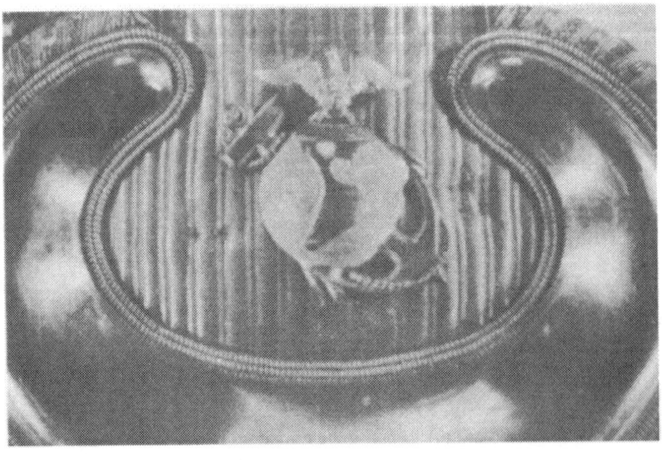

Figure 23. -Officer's epaulette ornament (right), 1892-1904.

Figure 24. - First Lieutenant John s. Bates, July 1, 1899, is shown in the undress coat prescribed by the 1892 Uniform Regulations.

Figure 25. - Enlisted black helmet Corps device, 1892-1904. Type (1) consisting of a slightly different pattern from Type (2) and containing on the back of the device a screw post fitted with a milled nut for securing it to the helmet.

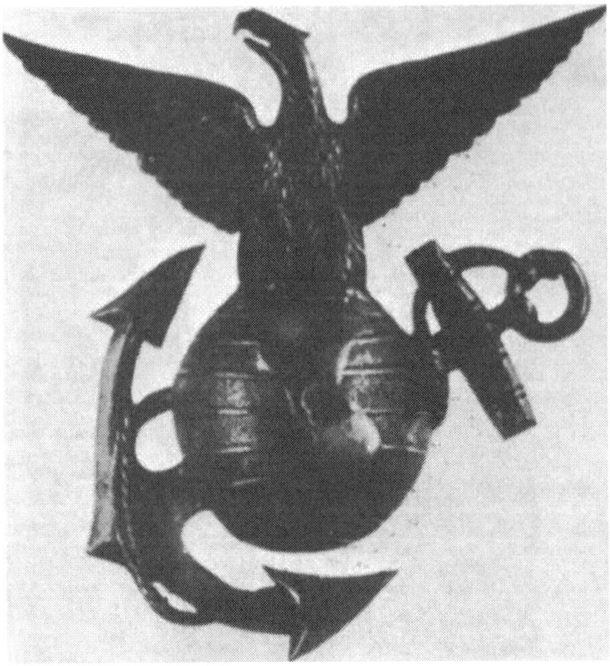

Figure 26. - Enlisted black helmet Corps device, 1892-1904.
Type (2) consisting of a slightly different pattern from Type (1) and containing on the back of the device two br ass-prong fasteners for securing the device to the helmet.

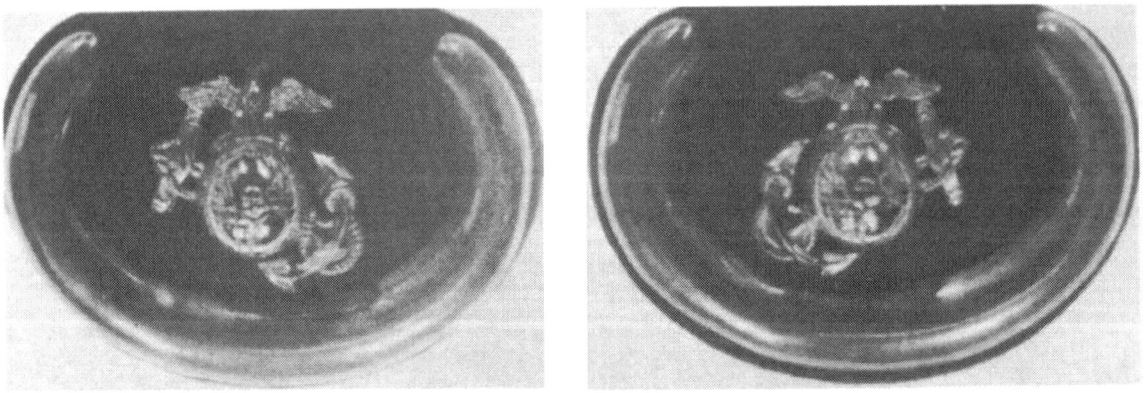

Figure 27. - Enlisted shoulder knot with Corps device, 1892-1904.

Figure 28. - Enlisted Marine shown in the full dress coat worn from 1892 to 1904.

Figure 29. - First Lieutenant Austin R. Davis, circa. 1898, in undress uniform. While the coat worn was prescribed in the 1892 Uniform Regulations, the undress cap was of the pattern first prescribed for wear in 1897.

Figure 30. - Group of Marine Officers, 1st Battalion Marines, 1898, in campaign unifor m,

While three off icers are wearing the undress cap ornament on their campaign hat on the lef t side, one officer is wearing his in the front of the hat, and one does not appear to be wearing any ornament

Figure 3 1.-Marines in the Peki°? relief expedition, 1900, earing the undress unifor m and campaign hat with ornaments on the left side of the hat.

Figure 32. - Enlisted Marines, circa 1900, in the Philippines, wearing the old blue field kit for tropical use. It should be noted that the undress cap ornaments are shown placed in the front of the campaign hat.

Figure 33. - Staff Officer's shoulder knot ornament (right), 1900-1904.

Figure 34. - Enlisted cap ornament, 1900-1908.

## CHAPTER IV

On 16 March 1904 another set of Uniform Regulations was approved. As in the past, the new regulations prescribed the various ornaments to be worn. For officers, ornaments were required for wear on the caps and hat, the chapeau, the coats, the mess jackets, the epaulettes for the Commandant and line officers, and the shoulder knots for staff officers.

The ornament prescribed for the officers' full-dress, undress and white caps was of slightly larger dimensions than those of the cap ornament described in the 1892 and 1900 Uniform Regulations. (See Appendix F-2)

The ornament for the Brigadier General Commandant's undress hat continued to be placed within its "wreath of gold acorn leaves" while his chapeau and those of staff officers displayed "a gold and burnished Marine Corps cap ornament."

For the front of the khaki cap and field hat, the new regulations provided "A device similar in all respects as to shape, size, and design to that (of the Corps device for caps), but made in dull-finished bronze metal." (Figs. 35 & 36)

The Corps device prescribed for the collars of the officer's undress coats (blue and white) and the white mess jacket changed from overall silver to a gold and silver combination and was of larger dimensions than its predecessor. (Fig. 37 & Appendix F-4)

At the same time the ornament for the collar of the officer's field (campaign) coat was made, "A device similar in all respects as to shape, size, and design to that for collars of undress coats, but made of dull-finished bronze metal":( Figs 38 & 39)

For the officer's epaulette, the 1904 Uniform Regulations prescribed that the "Corps device in silver" would be placed with the crescent. (Appendix F-6)

The device for the staff officer's shoulder knots was "to be worn on end of knot, flukes of anchor pointing toward the front," and was: "to be all silver and of same size as collar ornament (for undress coat)." This device as well as others, was to conform to standard samples kept in the office of the Quartermaster of the Marine Corps.

The ornaments for the enlisted men were to be worn on the full-dress cap, the dress (blue and white) caps and the field hat. For the enlisted caps the ornament was to be: "A Corps device of gilder's metal No. 14, of the same pattern and dimensions as the present standard sample in the Quartermaster's Department• ••" (Figs. 34 & 40)

For the enlisted field hat, "the cap ornament, together with battalion and company figures and letters, also in dull finished bronze, will be worn in front• ••"(Fig. 41)

Shortly after the 1904 Uniform Regulations were issued, an addenda was published containing slight modifications to some items of officers uniforms. Among these was a change involving the wear of the ornament on the undress coat worn by the Commandant. Heretofore, the Corps device had been worn on the coat collar. The modification contained in the addenda required: "The Corps device, as per pattern, to be worn in the center of the (shoulder) strap, the flukes of the anchors to point toward the front." No provision. was made for any device to replace the emblem on the coat collar.

In 1908, new uniform regulations were published containing all of those of 1904 as modified by its addenda, as well as a few additional changes that were made at the time of printing.

A comparison of the 1908 Uniform Regulations with those of 1904 modified by addenda, discloses no major changes involving the ornaments or the uniform parts on which they were worn.

Some minor points, however , were clarified by these regulations, for example: the prescribing of an eyelet in the front of the enlisted white cap cover to receive the post of the ornament as well as specifying that the ornament for the enlisted field hat should be, "A Corps device of the same pattern and dimensions in dull-finished bronze" as the cap ornament.

At this point the design of the ornaments was fairly well stabilized and re- mained so until the publication of the 1912 Uniform Regulations. Some small changes in uniform design occurred af ter the publication of the 1908 B.egulations however , most significantly, the introduction of a new style officer's undress coat in 1909 which was substantially the forerunner of the pr esent dress blue coat. The device worn on this garment was unchanged from that wor n on its predecessor and retained its place on the collar.

Additionally , a khaki cap cover was prescribed for enlisted men on 6 January 1911. The bronze ornament used with the field hat was to be worn when this cap cover was placed over the undress cap.

Figure 35. -Officer's service cap and field hat ornament, 1904-1914.

Figure 36. Officers of the Second Regiment, USMC, Cuba, September 1906, shown in the field uniform of the period.

Figure 37. - Captain W. H. Pritchett, circa. 1909, in blue undress uniform.

Figure 38. - Officer's service collar ornament (right). 1904-1914.

Figure 39. - Colonel Franklin J. Moses, 1910, shown wearing the field coat and khaki cap of the period,

Figure 40, - Enlisted Marines, circa. 1910, shown in the undress blue uniform prescribed for wear from 1904 to 1914.

Figure 41. - Enlisted Marine, circa. 1910, in the field uniform of the period.

# CHAPTER V

The 1912 Uniform Regulations introduced the predecessor of the familiar "green" uniform. Also the new Regulations began the period during which the forerunner of the modern enlisted "dress blue" uniform was first issued. Moreover, this period saw the garrison (overseas) cap and "Montana peak" field hat added to the Marines apparel.

With the appearance of the new uniforms in the Marine wardrobe, the enlisted full-dress coat and cap disappeared together with the white helmet that had been worn by officers and enlisted men since 1881. The Corps device worn on the staff officer's shoulder knot also had disappeared by this time.

In his directive dated 29 November 1912 accompanying the new regulations, Major General Commandant William P. Biddle notified all concerned that the new regulations applicable to officer's uniforms were to be complied with as soon as practicable, but not later than 1 January 1914. The directive also stated that: "The issue of the overcoat and winter field uniform to the enlisted men• • • • and the necessary alterations in their dress uniforms will be made as soon as practicable." In actuality the distinctive winter field green uniform was not worn until 1914 due to difficulty in procuring suitable material of the right shade of green.

When the 1912 Uniform Regulations were published, photographs and illustrations of uniforms and insignia were provided with the text, and this practice has continued to the present. The last illustrated regulations before 1912 were those of 1875.

For officers, the new regulations required that ornaments be worn on the caps and hat, the chapeau, the coats, and the epaulets.

A description of the ornament required for the officer's full-dress, undress and white caps as well as the special full-dress and dress caps of the Leader of the band can be found in Appendix G-2. (Figs. 42 & 43)

The ornament for the full-dress cap of the Commandant was to be placed on the front of the cap band within a "wreath of oak leaves embroidered in gold."

On the chapeau of the Commandant and staff officers, the ornament worn was: ."A Corps device of gold of the same size as the device prescribed for caps." (Fig. 44)

For the officer's summer field cap, the winter field cap and field hat the device worn was: "similar in all respects as to shape, size, and design to that prescribed for full-dress and undress cap, and shall be of dull-finish bronze metal." (Figs. 45 & 46)

The ornaments required to be worn on the officer's undress and white undress coats, the white mess jacket and on the special full-dress and dress coats of the Leader of the Band was much the same as that worn on full-dress caps etc. differing primarily in size and lacking the anchor cable or rope. (Appendix G-4 and Figs. 47, 48 & 49)

For the collars of officer's summer and winter field coats, the reg,lations required that the Corps devices be: "similar in all respects as to shape, size, and design to that prescribed for the collars of undress coats, and shall be made of dull-finish bronze metal." (Fig. 50)

The ornament worn on the officer's epaulets were to be all silver. (Appendix G-6 and Figs. 51 & 52)

Corps devices were worn by enlisted men on the caps and hat. The description of the ornament worn on the enlisted dress cap (with or without white cap cover) as well as the special full-dress, and full-dress caps of the enlisted members of the band was:

> "A Corps device, struck from one piece .of gilder's metal No. 14, of the same pattern and dimensions as the standard sample in the Quartermaster's Departmen t of the Marine Corps• • • •" (Figs. 53, 54 & 55)

For the enlisted summer field cap (which was no more than the dress cap with a khaki cap cover) , the winter field cap, and the field hat the prescribed device was "a Corps device of the same pattern and dimensions" as the dress cap in dull- finish bronze. (Figs. 56, 57 & 58)

After the issue of the 1912 Regulations, changes appeared periodically that ·modified or replaced certain uniform items. The changes which substantially aff ected articles of clothing, on which the ornaments were worn were:

> Changes No. 1 of 3 February 1915 abolished the white cap cover for enlisted member of the band, and substituted a white cap in its place.
>
> Changes No. 4 of 26 December 1916 prescribed the following uniforms for warrant officers:
>
> Undress
> White undress
> Summer field dress
> Winter field dress
>
> Changes No. 8 of 12 May 1917 provided a cap frame for the enlisted men upon which the white and summer field cap khaki covers were to be worn.

In 1917 new regulations were published containing all of those of 1912 as modified by Changes 1 through 8. This new revision was also altered by additional changes, and the following paragraphs enumerate those that introduced new ornaments to be worn, or materially aff ected the items of clothing on which the ornaments were worn:

> Changes No. 9 of 14 September 1917, No. 10 of 26 October 1917 and No. 12 of 31 January 1918 contained slight modifications to the officer's winter field coat.
>
> In addition, Changes No. 12 modified the officer's undress coat to a limited degree while Changes No. 13 of 29 March 1918 clarified the specifications of the officer's summer field hat.

Also during this period which encompassed the First World War, changes other than those specified in the uniform regulations took place.

On 14 June 1917 the first contingent of Marines sailed for France taking with them modified versions of their distinctive summer and winter field uniforms. Later contingents leaving for France also took their khaki and "field green" uni- forms with them.

Due to supply problems, a memor andum from the Adjutant Gener al, Head- quarters American Expeditionary Forces was issued on 10 January 1918 stating in part: "During the present war, the regulation Army uniform will be regularly issued to the Marines of the A.E.F., therebygr adually replacing the Marine uniform" ••"

In compliance with this order , the Army uniform was then issued and worn by all Marines in the A.E.F.

Consequently , it became dif ficult to distinguish between the enlisted men of the Army and those of the Marine Corps. The opportunity to do something about this arose on 5 August 1918, when Assistant Secretary of the Navy, Franklin D. Roosevelt inspected the Fourth Brigade near the Marbache Sector. In Appendix C of the Annual Repor t of the Secretary of the Navy for the Fiscal Year 1918, Assistant Secretary Roosevelt reported:

> ". • • • The Marine themselves were near Nancy , and I inspected them there, and took occasion to give some slight token of our appreciation by directing that every man in the brigade be allowed to wear the Marine Corps device on the collar points of his shirt. At that time Maj. Gen. J. A. Lejeune, United States Marine Corps, was in command of the Second Division, American Expeditionary Force, and Brig. Gen. W. C. Neville, United States Marine Corps, in command of the Marine Brigade."

With this authoriz ation in hand, Gener al Neville directed the Marine Paymaster in Paris to secure a supply of devices. Instead of securing a smaller though identical version of the officer's collar device for field coats, the Paymaster procured the Army type collar disk with the Marine Corps emblem stamped on it. **(Fig. 59)**

Although these devices were for use only with the Army uniform in France, their introduction represented the first time that all enlisted men in a given area wore a collar ornament.*·

*Enlisted members of the band wore small white metal Corps collar devices from **1892** to 1900.

During their service in France, the Marines were also to acquire the now **familiar** "overseas cap" that was prescribed for wear by the Army under Head- quarters A.E.F. General Orders No. 7 of 9 January 1918. Contemporary photographs of the period show the Corps device being worn on the left front side of the cap, with officers using either the bronze cap and hat ornament or the lef t field collar device and enlisted men using the Corps device from the field hat. (Fig. 60)

Photographs of the period also show Marines in France wearing the field hat Corps device on the front of the steel helmet. Although no documentation can be found authorizing its use, the Sixth Marine General Order No. 5 of 29 January 1919 directed that the Marine Corps Emblem would not "be affixed to helmets" until further orders.

Other items of uniform on which the Corps device was worn during this period were the service caps and field hats of Women enrolled in the Marine Corps Reserve. The caps and hats authorized for women marines by Marine Corps General Order No. 62 published in December 1918 were:

1 Cap, Glengarry, summer field
1 Cap, Glengarry, winter field
1 Hat, field

The Glengarry cap was an "overseas cap". The summer field model was made of khaki, while the winter field model was manufactured of olive drab material.

The ornament worn on these caps and hat was the enlisted bronze cap and hat Corps device. The device was worn on the left front side of the Glengarry cap and on the front of the field hat. (Figs. 61 & 62)

After the War, additional changes were instituted involving the ornaments and the items of uniforms on which they were worn.

Change No. 18 dated 7 October 1919 eliminated the Special full-dress and full-dress uniforms (except for members of the band), which included the chapeau, the full-dress cap and the epaulet. The mess jacket and white mess jackets were eliminated and replaced by an evening dress jacket of dark blue cloth. The dress collar ornaments were to be worn on the collar of this jacket.

Collar ornaments were finally authorized for enlisted men by Changes No. 20 of 25 February 1920. These were similar to those worn by officers but were of smaller dimensions and were of gilt for use on the dress coat and dull-finish bronze for the summer and winter service coats. (Appendix G-8 and Figs. 63 & 64)

With the publishing of Changes No. 22 dated 9 October 1920, the garrison (overseas) cap became an official item of the Marine Corps uniform. At this time only the summer service garrison cap was prescribed, while the authorization of the winter service cap remained in the future. The ornament to be used on this cap was the dull-finish Corps device worn on the field caps and hat.

On 29 October 1921, Major General Commandant John A. Lejeune, issued a Circular Letter informing all officers that the mess jacket as part of the evening dress uniform had been restored with certain modifications. The collar had been changed and the dress collar Corps devices were removed from it. In addition, shoulder knots with rank insignia and dress collar Corps devices were prescribed for wear with this uniform.

The Major General Commandant's Circular Letter of 24 April 1922 again listed the specifications for the evening dress uniform as well as directing that it had to be acquired. by 24 April 1922. The letter also gave the specifications for the new dress, undress and service caps of officers, but stated that the new regulations to be published would permit the present regulations cap to be worn until no longer serviceable. The new caps described in this letter were very similar in shape to those worn today.

Figure 42. - Officer's full-dress and undress cap ornament, 1914-1919; Officer's dress cap ornament, 1919-1937.

Figl.ire 43. - Officer's full-dress and undress cap ornament, 1914-1919; Officer's dress cap ornament, 1919-1937.

Figure 44. - Commandant's and staff officer's chapeau ornament, 1914-1919.

Figure 45. -Officers' field cap and hat ornament, 1914-1922; Officer bronze cap and hat ornament, 1922-1930: Officer's service cap and hat ornament, 1930-1937.

Figure 46. -Officer in the winter field uniform first worn in 1914.

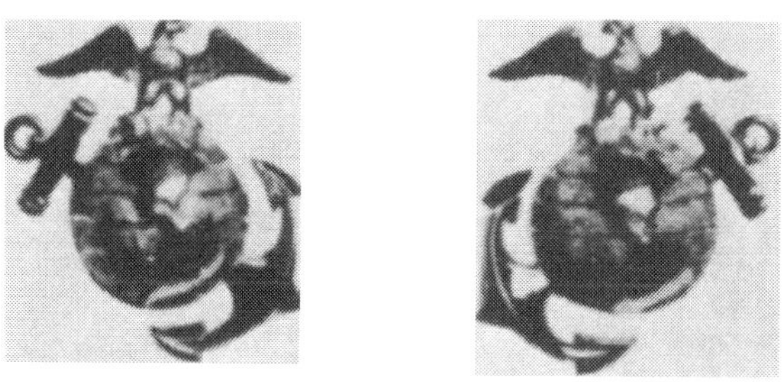

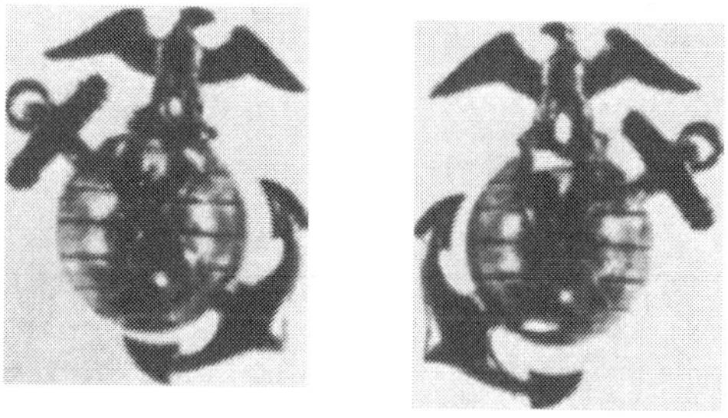

Figure 47. - Officer's undress collar ornaments, 1914-1919 ; officer's dress collar ornaments, 1919-1922; officer's dress collar , epaulet and shoulder knot ornaments, 1922-1942 ; officer's dress collar and shoulder knot ornaments, 1942-1948; officer's dress collar ornaments, 1948-1962.

Figure 48. - Officer's undress collar ornaments, 1914-1919 ; officer's dress collar ornaments, 1919-1922 ; officer's dress collar , epaulet and shoulder knot ornaments, 1922-1942 ; officer's dress collarand shoulder knot ornaments 1942-1948; officer's dress collar ornaments, 1948-1962.

Figure 49. - Second Lieutenant Harry A. Ellsworth, 1917, in white undress uniform.

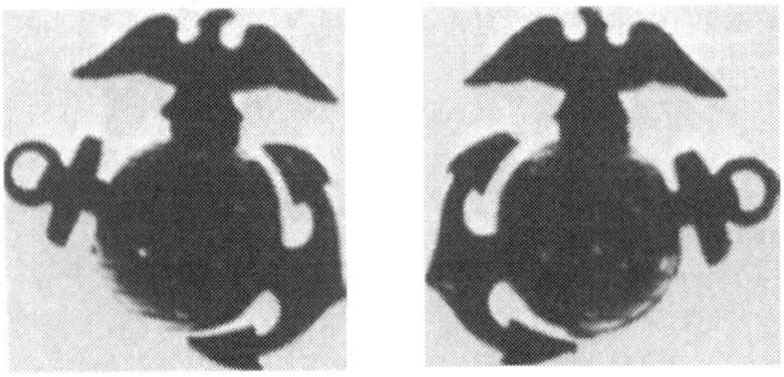

Figure 50. - Officer's field coat collar ornaments, 1914-1922; officer's bronze collar ornaments, 1922-1927.

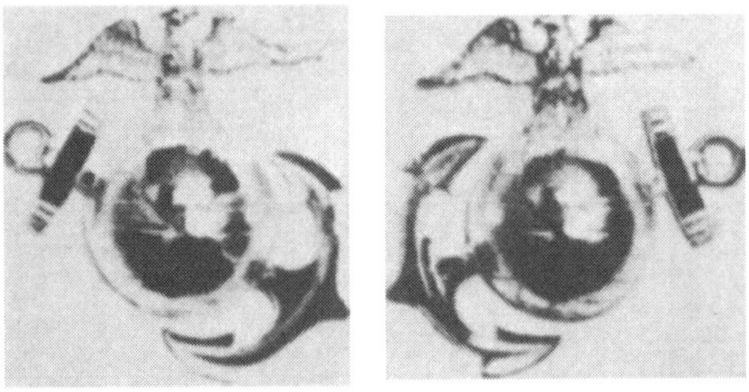

Figure 51. - Officer's epaulet ornaments, 1914-1919.

Figure 52. - Colonel Harry K. White, 1916, in full-dress coat with epaulets.

Figure 53. - Enlisted dress cap ornament, 1914-1922,

Figure 54. - Enlisted dress cap ornament, 1914-1922.

Figure 55. - Gunnery Sergeant Daniel Daly, circa. 1919, in undress uniform.

Figure 56. - Enlisted field cap and hat ornament, 1914-1922.

Figure 57. - Enlisted field cap and hat ornament. 1914-1922.

Figure 58. - Sergeant George Schifferer is shown in the winter field coat worn from 1914 to 1928. The "Montana Peak" field hat he is wearing was authorized for general use from 1914 to 1942, and again from 1961 to the present, but for Drill Instructors at Recruit Depots and Rifle Range personnel only.

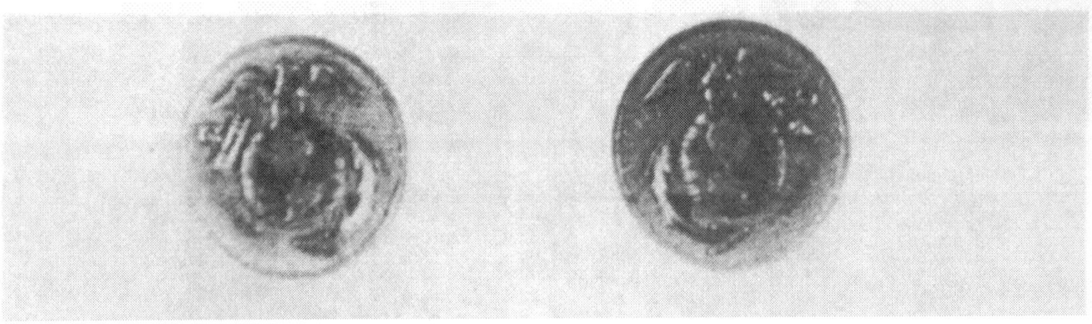

Figure 59. - Enlisted service collar ornaments, 1918-1919.

L'igure 60. - Brigadier Gener al Smedley D. Butler in France, 1918-1919, in winter field uniform with overseas cap.

Figu re 61. - Private Carrie E. Kenny , in winter field unifor m with field hat worn from 1918-1919.

Figure 62. - Private Carrie E. Kenny, in winter field uniform with winter field Glengarry cap. Women were enlisted in the M arine Corps Reserve during the World War I period from 12 August 1918 through 30 July 1919.

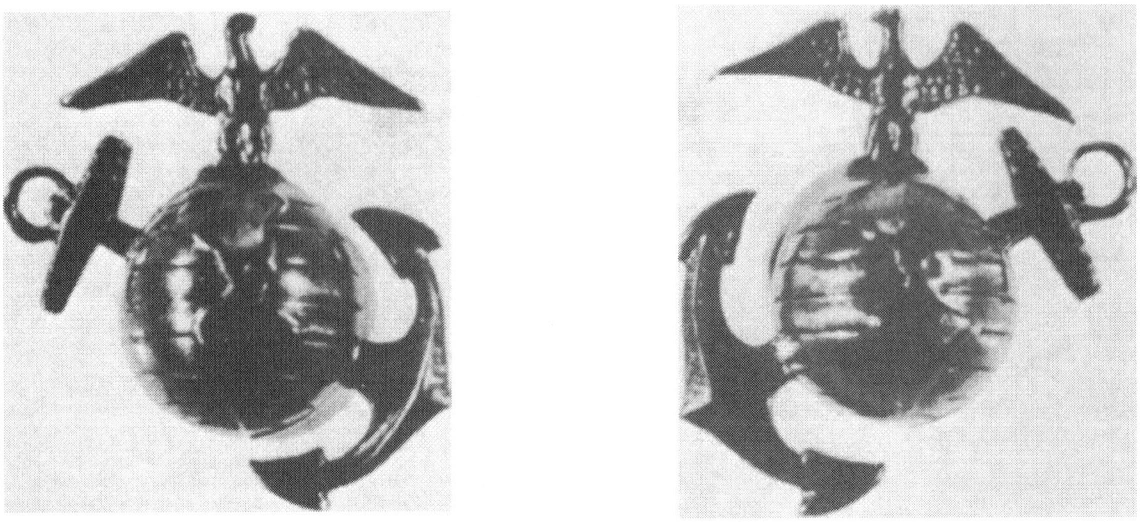

Figure 63. -Enlisted dress coat collar ornaments, 1920-1930.

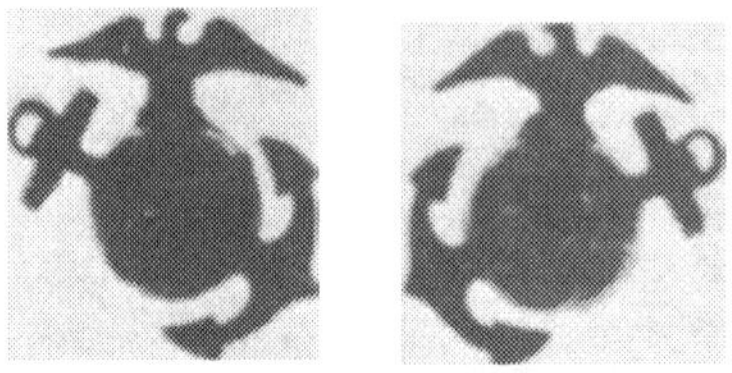

Figure 64. -Enlisted service coat collar ornaments, 1920-1930.

# CHAPTER VI

In 1922 yet another set of Uniform Regulations was issued. This was published on 11 September of that year and superceded the regulations of 1917 with the numerous modifications and changes but the ornaments prescribed for officers were the same as those which had been worn continuously since 1912. The new enlisted cap and hat ornaments were very similar to those dating back to that year, while their collar ornaments were the same as those described in Uniform Regulation Changes No. 20 of 25 February 1920. The new publication also contained a few additional changes that had not previously appeared. The most notable of these changes involving an article of uniform on which the ornament was worn was the abolition of the enlisted bell-crown caps and their replacement with caps similar in shape to those worn today. ( Fig. 65) This had previously been authorized for officers by the Major General Commandant's Circular Letter of 24 April 1922. Also, the visors of all service caps were changed from tan to a russet (nut brown) color, and officers were provided a field hat similar to the recently modified enlisted field hat.

Further, the sections covering the placement of Corps devices on officer's coat collars were amended to read: "Center of ornament 1 1/2 inches from the neck opening" with staff department insignia placed "to the rear of the collar ornament, 2 inches center of ornament to center of insignia".

The proper ornaments to be worn on the officer's caps and hats were provided for in the new regulations as follows:

> "The gold and silver cap ornament shall be worn on the dress and undress white caps, and the bronze cap and hat ornament on the winter and summer service caps, the garrison cap, and the field hat, being worn on the garrison the cap on left front side, with the center of the ornament 2 inches fr.om the ( Figs. front." 42, 43, 45 & 66)

The gold and silver ornament was also worn ori the full-dress and special full-dress caps of the Leader of the Band. While on the dress cap of the Commandant the ornament was to be placed within a wreath of gold embroidered oak leaves. Fig. 67 Specifications for these ornaments are contained in Appendixes H-2, & **H-3.**

The gold and silver collar ornaments were worn on the officer's dress and undress white coats as well as shoulder knots. See Appendix H-5 and Figures 47 & 48. The bronze collar ornaments were worn on the officer's winter and summer service coats. See Appendix H-6 and Figure 50.

Ornaments prescribed for the enlisted men were for both the caps and hat and for the collars of the coats. For the caps and hat, the regulations provided that:

> "The gilt cap ornament shall be worn by all enlisted men on the dress and white caps, and the bronze cap and hat ornament on the winter and summer service caps, the garrison cap, and the field hat. On the garrison cap, the bronze cap and the hat ornament shall be worn on the left front side, in the place provided therefor."

The gilt cap ornament was also worn on the full-dress and special full-dress cap of members of the band, (Figs. 68, 69 & 70)

For the coats of enlisted nen, the regulations provided:

> "Gilt collar ornaments are presc ribed for the collar of the dress coat, and bronze collar ornaments for the winter and summer service coats." (Fig, 71)

In addition, the gilt collar ornaments were worn on the collars of the special full-dress and full-dress coats of the Band, (Figs. 63, 64 & 72)

Altogether , three printed changes and a number of Circular Letters were published to modify the 1922 Regulations as initially issued,

In Changes No. 1 dated 22 July 1924, the wreath of gold embroidered oak leaves was eliminated from the band of the Commandant's dress cap; the gilt collar orna- ments were removed from the special full-dress coat of Marine Bandsmen; and the dimensions for the caps were modified slightly,

Except for very minor corrections the main purpose of Changes No. 2 dated 6 September 192 4 together with its accompanying memorandum of 15 August 1925, was to provide the illustration plates that were missing at the time the regulations were initially published,

On 14 April 192 6 Circular Letters Nos, 16 and 17 were published presc ribing roll-collar service coats for both officers and enlisted men,

The dates these coats were to be obligatory was 1 July 1927 for winter service and 1 J anuary 1928 for summer service, but apparently due to procu rement difficulties the time for acquisition of the summer service coat for enlisted men was extended

until 15 August 1928 for posts in the U.S. (outside the U. S, - until the supply of coats standing collar was exhausted) , with the coat not to be worn prior to 15 The time October 1928, was similarly extended for enlisted winter service coat until 1 October 192 8. (Fig. 73)

The introduction of roll-collar coats brought about the need for smaller orna- ments. Circular Letter No, 46 of 6 December 192 6 specified that:

> "A small bronze collar ornament has been adopted for roll-collar coats for officers, to be obligatory on 1 July 192 7 , for winter service, and 1 J anuary 1928 for summer service, but optional in the meantime, These ornaments will be three-fou rths the size of the bronz e collar ornaments heretofore specified, and will be made in pairs, right and lef t, in one piece and will follow the design of the basic Marine Corps emblem, omitting the motto, ribbon and rope."
>
> ". , , • The ornaments will be centered on the collar between the crease roll and the outer edgc, ccnter of hemisphere 1 inch above the horizontal edge of notch, and the insignia will be placed on the lapel with the center mid-way on a line bisec ting the angle of the peak of lapel. The manner of placing these devices is illustrated by the inclosed $ketch", (Fig. 74)

The sketch accompanying Circular Letter No. 46 introduced another style of ornamen t, diff ereing both in size and shape from all others previously authorized. Its design was taken from the STANDAR D G. S. I\IARIK E COR PS EI\IBLE I\I , approved by :\I ajor Gener al Commandant John A. Lejeune and Ac ting Secretary of the Navy Theodore D. Robin son on 28 :\lay 1925. The E:\ IBLE .I\ I was drawn up by Staff Sergeant Joseph H. Burnett ofthe Quartermaster's Office , af ter a year of research. (Fig. 75)

DeriYation of the uniform ornament from the STANDAR D U. S. I\ IARIN E CORPS L\IBLE:\I establishes a trend for the future. As will be seen' in succeeding chapters the Standard E mblem was to undergo two more changes and shortly af ter each modif ication a new ornament was presc ribed.

Figure 65. - Marine Corporal James M. McCarthy, circa, 1922, in dress uniform.

Figure 66. - General Officer, 1922, in evening dress uniform.

Figure 67. - Major General Commandant John A. Lejeune, circa. 1922-1924, in dress uniform with overcoat. His dress cap contains "the gold and silver ornament encircled by a wreath of oak leaves embroidered in gold."

Figure 68. - Enlisted dress cap ornament, 1922-1930.

Figure. 69. -Enlisted dress cap ornament, 1922-930.

Figure 70. - Enlisted bronze cap and hat ornament, 1922-1930.

Figure 71. - Sergeant Howard A. Drapeau, circa. 1922 in winter field uniform with field hat.

Figure 72, - Second Leader , United States Marine Band , 1922, in full dress uniform

Figure 73, - Enlisted Marines ofthe 6th Battery , 10th Artillery and the 80th Co. 6th Regt., 3rd Brigade in winter service uniform (with roll-collar) , leave for home 21 Nov. 1928 - Tientsin, China.

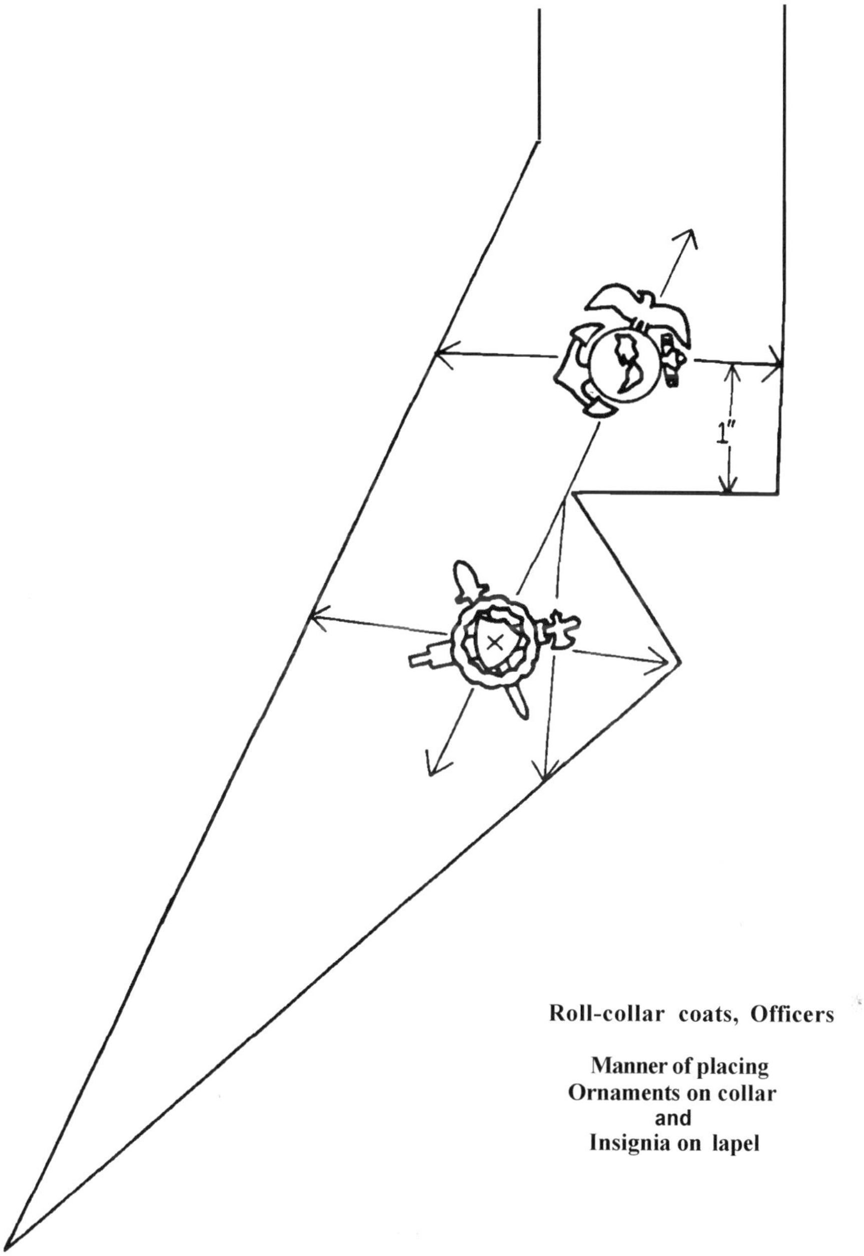


Figure 74. - Illustration accompanying Circular Letter No. 46 of 6 December 1926.

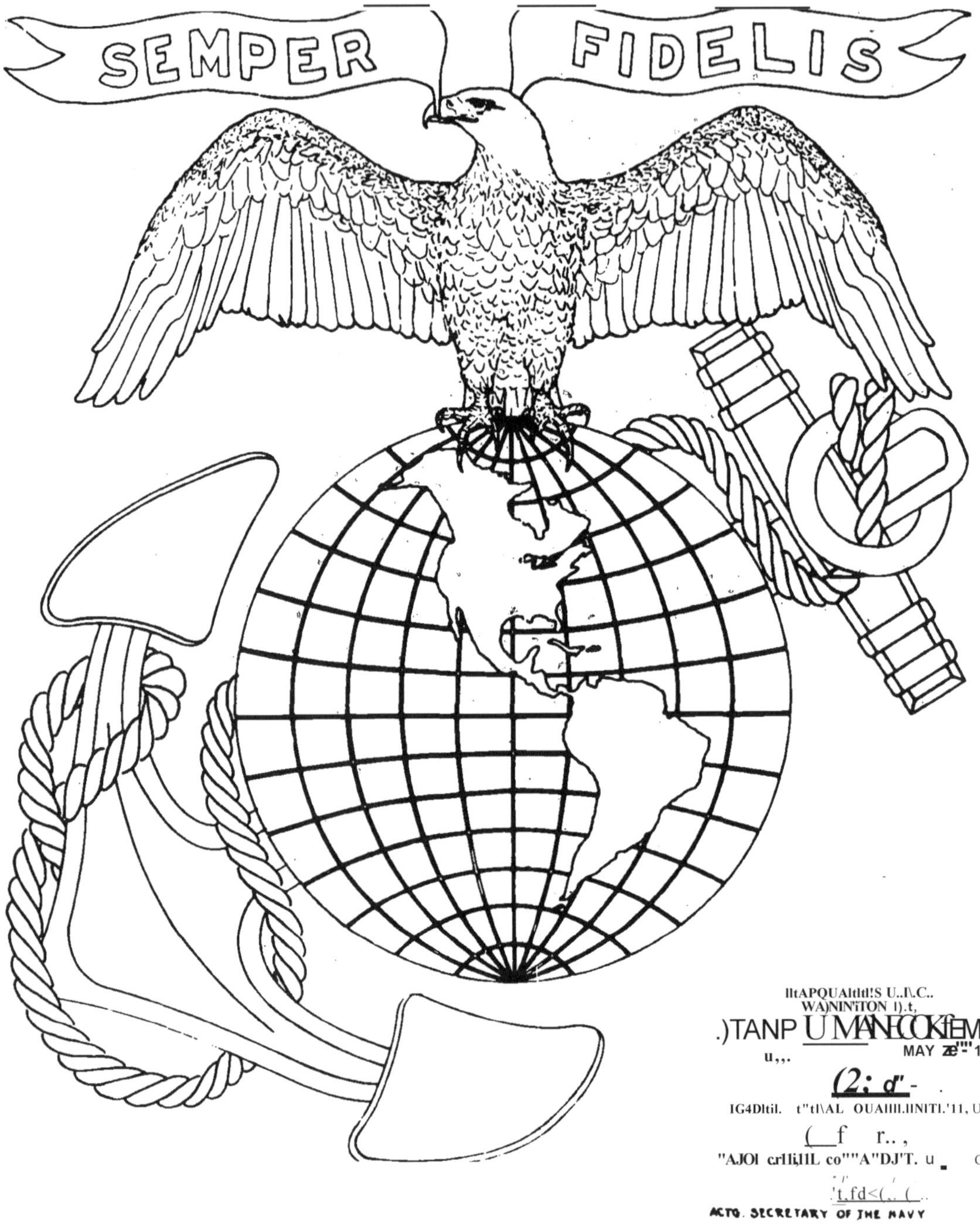


Figure 75. - Standard U. S, Marine Corps Emblem , approved 28 May 1925,

# CHAPTER VII

In 1929 another set of Uniform Regulations was prepared for publication. These regulations, issued on 22 April 1930 , were a composite of those of 1922 as modified by its changes and Circular Letters, plus a few additional changes that appeared for the first time.

Among the new changes was the modification in the design of the enlisted dress and service collar ornaments to conform to the STANDARD U. S. MARINE CORPS EMBLEM., adopted in 1925. This change in the officer's service collar ornaments was previously authorized by Changes No. 46 of 6 December 1926.

Another new change involving the ornaments and the uniform on which they were worn, prescribed a special full-dress uniform with the full-dress cap and shoulder knots for officers detailed to duty at the White House.

Other additions to the new regulations were the appearance of both the winter service garrison (overseas) cap for officers and enlisted men and the officer's white mess·jacket.

An examination of the ornaments worn during this period, indicates that while the officer's cap and hat ornament was virtually the same as that worn since 1914, the enlisted cap and hat ornament was modified slightly in design from its predecessor. (Figs. 79 & 80)

Under the new regulations, the officers dress cap ornament was to be worn on the dress and white undress caps, as well as the full-dress and special full-dress cap for officers detailed for White House duty and for the Leader of the Marine Band. (Figs. 42 & 43) Specifications for this ornament are contained in Appendix 1-2.

The officers service cap and hat ornament was worn on the summer service and winter service caps, the summer service and winter service garrison caps, and on the field hat. (Fig. 45) See Appendix 1-3 for pertinent details of this ornament.

The dress collar and shoulder knot ornaments were worn on the officers dress and undress white coats, as well as the white mess jacket and shoulder knots. Specifications for these ornaments are listed in Appendix 1-5. (Figs. 47, 48 & 76)

For the officers summer service and winter service coats, the service collar ornaments were prescribed. (See Appendix 1-6 and Figs. 77 & 78.)

For the enlisted cap and hat ornaments the regulations provided that: "The gilt ornament (Fig. 79) shall be worn by all enlisted men on the dress and white caps, and the bronze cap and hat ornament (Fig. 80) on the summer service and winter service caps, the (summer service and winter service) garrison caps, and the field hat. • • •" The gilt cap ornament was also worn on full-dress and special full-dress caps of the Marine Bandsmen.

The collar ornaments authorized for the enlisted men were ofboth gilt and bronze. The gilt collar ornaments (Fig. 81) were worn on the dress coat while the bronze ornaments (Fig. 82) were worn on the summer service and winter service coats. The gilt collar ornaments were also worn on the full dress coats of Marine Bandsmen.

Before the 1929 Regulations were replaced by those of 1937, severa.l changes took place involving articles of uniform on which the ornaments were worn.

One significant change appeared in Changes No. 1of 30 June 1932. The shoulder knots worn on the full dress coats of the Aide-de-camp to the President and officers detailed for duty at the White House "were discarded and were replaced by epaulets." This resulted in another type ornament being authorized that was to be worn on the epaulet only. See Appendix I-9d.

Another change was published on 31 August 1935 authorizing a new white mess jacket that was to be obligatory on 1 July 1936. The placement of the ornaments was on the shoulder straps, as opposed to the collar where they were previously worn.

Figure n. - Captai n John Kaluf, 1936, in blue undress B un iform.

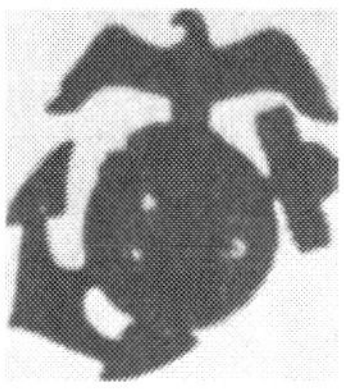

Figu re 77. - Officer's service collar orn amen t (lef t), 1927-1937.

figu rt 78. - Lt. Colonel Cli fton B. Cates , circa. 193C, in win ter service C un ifor m.

Figure 79. - Enlisted gilt cap ornament, 1930-1937.

Figure 80. - Enlisted bronze cap and hat ornament, 1930-1937.

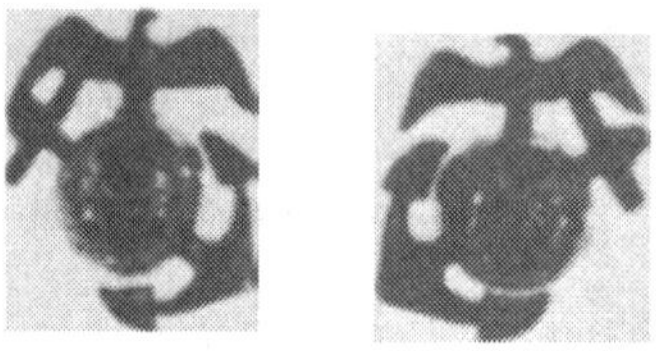

Figure 81. - Enlisted gilt collar ornaments, 1930-1937.

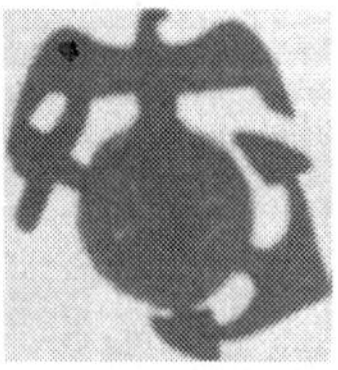

Figure 82. - Enlisted bronze collar ornament, (right ornament) 1930-1937.

# CHAPTER VIII

The Unifor m Regulations that were to see the Marines through World War II and into the post-war era were prescribed in 1937. During this period, some of the more important developments involving ornaments and uniforms included experimentation with plastic cap and collar ornaments, the adoption of wool green and khaki service jackets, the modification of the enlisted dress blue coat, and the authorization of a new officer's evening dress uniform.

These regulations, issued on 21 M ay 1937, were as usual a combination of the preceeding (1929) edition with all of its modifications, as well as containing several new changes that appeared for the first time.

On 16 March 1936 a new STANDARD U. S. MARINE CORPS EMBLEM was approved by Major General Commandant John H. Russell and Secretary of the Navy Claude A. Swanson. ( Fig. 84) This change was reflected in the design of the orna- ment that appeared in the 1937 Regulations the following year.

Due to the time needed by the insignia manufacturers to re-tool, the new ornaments were not seen immediately. When the officers' ornaments finally appeared, several diff erent styles were noticed. This was due to the fact that although certain required specifications were provided, there was no standard pattern designated, and each manufacturer produced an ornament that differed slightly in design from those of other makers. (As examples see Figs. 85, 86 & 87)

In examining officer's cap ornaments manufactured during this period, it can be seen that the anchor rope or cable was wrapped around the flukes of the anchor. Prior to this the rope was arranged in a manner that it passed over the base of the top anchor fluke.

Moreover , it should be noted that most officer's dress and service ornaments produced during World War II were made of sterling silver. This came about because of the scarcity of the other metals that were needed for the war effort.

When the enlisted ornaments issued through the supply system appeared, uni- formity in design had been achieved. This resulted from efforts by the Supply Department to furnish standard ornament hubs to manuf acturers who had con- tracted to produce ornaments.

One significant change involving uniform items on which ornaments were worn appeared in the new 1937 Regulatibns. This was the elimination of the full-dress and special full-dress cap for the enlisted members of the Marine Band.

Following the example of its predecessors, the 1937 Regulations furnished the specifications for all officer's ornaments, including instructions as to how they were to be worn. For the cap and hat ornaments it provided that: "The dress cap orna- ment shall be worn on the full-dress, dress, and whi,te caps; ( Figs. 85, 86 & 87) , and the service cap and hat ornament on the winter and summer service caps, the winter and summer service garrison caps, and the field hat." ( Figs. 88, ·89 & 90) See Appendix J-2 & J-3 for the specifications of these ornaments.

The requirements for the officer's collar ornaments, were that: "The dress collar ornaments shall be worn on the dress and white coats. the white mess jacket. and on the epaulets and shoulder knots. ( Figs. 91, 92 & 93) The service collar ornaments shall be worn on the summer and winter service coats." ( Figs. 94 & 95) Appendix J-5 & J-6 contain the specifications of these ornaments.

Instructions for the wearing of the enlisted cap and hat ornament were that:

> "The gilt cap ornament shall be worn by all enlisted men on the dress and white caps. (Figs. 96 & 97), and the bronze cap and hat ornament on the winter and summer service caps. the garrison caps. and the field hat. ( Fig. 98) On the garrison caps. the bronze cap and hat ornaments shall be worn on the left front side, in the place provided therefor".

For the enlisted men. the gilt collar ornaments were to be worn on the dress coat as well as on the full-dress coat of the enlisted members of tile Band. ( Fig. 99) The bronze collar ornaments were to be worn on the winter and summer service coats. ( Figs. 100 & 101)

From the time of its issue. many important changes took place that extensively modified the 1937 Uniform Regulations.

On 6 May 1938. Circular Letter No. 242 was issued requiring that the left service collar ornament be substituted for the service cap and hat ornament on the summer and winter service garrison caps. This change has remained in force to the present day.

Circular Letter No. 246 of 31 May 1938 eliminated the enlisted summer service coat while Circular Letter No. 413 of 10 December 1940 prescribed the fiber helmet as an article of uniform to be worn. The service cap and hat ornament was worn on this helmet, as is indicated by an examination of contemporaneous photographs. (Fig. 102)

The field hat was abolished as an article of uniform by Circular Letter No. 584 of 28 May 1942. As a result Circular Letter No. 586 of 8 June 1942 the "service cap and hat ornament" was redesignated "service cap ornament" and the words "field hat" was striken from the description of uniform articles on which it was worn.

Sometime during the early part of World War II (possibly late 1942 or early 1943), because of the scarcity of certain critical metals, it was decided to issue enlisted service cap and collar ornaments made of dark-brown plastic. (Figs. 103 & 104)

During the time the First Marine Division was in Australia in 1943, convalescing from the ordeal of Guadalcanal, as a matter of expediency. the "Aussie" battle jacket was issued as a replacement for the winter service coat. Also due to supply difficulties, ornaments were procured locally as evidenced by the specimens shown in Figures 105 and 106. The ornament bears the marking "K. O. LUKE MEL- BOURNE."

Probably as a result of favorable reactions to the Australian battle jacket, an officers winter service (field) jacket closely resembling it was authorized by Letter of Instruction 754 of 20 May 1944. Letter of Instruction No. 885 of 15 November 1944 provided that this jacket would be optional and that ". . . collar ornaments will be worn in the same manner as prescribed for summer and winter service coats." In turn, Letter of Instruction No. 928 of 17 January 1945 the specifications for the jacket were slightly modified as well as including a provision for a 1/8" diameter worked eyelet" on its lapel for the collar ornaments. Further, in Letter of Instruction No. 1245 of 21 March 1946 the jacket was again slightly modified and was redesignated the Jacket, service, wool, green.

The last event concerning officers service jackets came about with the publication of Letter of Instruction No. 1274 of 9 May 1946 which authorized a khaki service jacket for officers. Subsequently the specifications for this jacket were contained in Letter of Instruction No. 1291 of 7 June 1946.

For the enlisted men both the wool green and khaki service jackets were adopted in 1946. The wool green jacket was adopted in February of that year while the khaki jacket appeared during that summer. (Figs. 107 & 108)

An additional uniform change came about by the decision to bring back the "dress blue" uniform whose wear had been discontinued during World War II. An article in the August 1946 issue of THE LEATHERNECK stated that:

> "It was . . .directed by the Commandant that September 1, 1946, be designated as the date when blue dress would once more become the uniform for shipboard wear, special formations and liberty. . ."

When the dress uniform did appear, its coat contained four outer pockets, similar to those on the winter service coat, and the tin wire grommet in the cap had been replaced by a half-inch wide metal strip. (Fig. 109)

One more change to be mentioned at this time occurred in 1948 with the publication of Marine Corps Bulletin 8-48 which announced a new evening dress uniform for officers. The main modifications included the removal of the shoulder knots (with dress ornaments) and the redesign of the collar with the placement of the dress ornaments in the appropriate place thereon.

## MARINE CORPS WOMEN'S RESERVE

On 13 February 1943 the Marine Corps Women's Reserve was established, representing the second time in Marine Corps history that women were enrolled within its ranks. The first instance was during World War I, when 305 women were enlisted to perform clerical jobs, mostly at Headquarters Marine Corps. Chapter V includes the pertinent uniform and emblem information regarding these early Women Marines.

Among the new organization's first set of regulations was a publication entitled "Uniform Regulations, U. S. Marine Corps Women's Reserve, 1943" which was enclosure (A) to Letter of Instruction No. 523 of 27 August 1943. These regulations were the result of the actions of a uniform board established on 17 June 1943, which submitted its recommendations to the Commandant, Lieutenant ueneral Thomas Holcomb, in July for his approval , and were subsequently approved by Secretary of the Navy Frank Knox on 25 August 1943.

Ornaments were prescribed in the regulations for both officers and enlisted women. The proper usuage of officers ornaments was:

CAP AND HAT ORNAMENTS

The service cap ornament shall be worn on the winter service cap, and dress cap ornament on summer dress and summer service caps.

COLLAR ORNAMENTS

( a) Service collar ornaments shall be worn on the winter service coat centered on the collar. . •

(b) Dress ornaments shall be worn on the summer dress and summer service coats, centered on the collar• • • (Fig. 110)

For the enlisted ornaments the regulations provided that:

CAP AND HAT ORNAMENTS

The dress (gilt) cap ornament shall be worn on the summer cap; and the service (bronze) cap ornament on summer service hat and winter service cap.

COLLAR ORNAMENTS

(a) ) Service (bronze) collar ornaments shall be worn on winter service coat. . . ( Fig. 111)

(b) Dress (gilt) collar ornaments shall be worn on summer dress coat, summer service B and summ er service C coats, and servie collar ornaments on summer service A coat. • ." (Fig. 112)

No specifications for these ornaments were given, but an examination of all .available material confirms that these were the same as those prescribed for male personnel of the period.

As in the case of all uniform regulations, these were subject to modification by various printed changes.

In letter of Instruction No. 670 of 24 February 1944 the officer's summer dress cap was eliminated, and .was replaced by the summer service cap. An officer's summer dress A coat was also added on which dress collar ornaments were to be worn "on the center line of each shoulder strap, 3/4 inch from armhole seam with head of eagle facing front." (Fig. 113)

Letter of Instruction No. 777 of 9 June 1944, contained several pertinent changes. A garrison cap of sage-green material trimmed with white braid was authorized, on which the left service collar ornament was worn. This cap, worn with the summer service A uniform, was optional for officers, but replaced the light green hat worn by enlisted women. The section of the regulations dealing with the method of wearing the officers cap ornaments was amended accordingly as well as the paragraphs concerning the wearing of the officrs collar ornament. See Appendix K-5b.

For the enlisted women, the section pertaining to the method of wearing the cap and hat ornament and the paragraph in the regulations dealing with the proper ornament to be worn on the enlisted summer coats were revised accordingly. See Appendix K-5c.

At about the time the garrisor cap was authorized, a utility cap ofthe same pattern, made of6ounce green twill, to be worn as part ofthe utility uniform also was approved. The left service collar ornament '#as worn in the front on the left side of this cap, in the same manner the ornament was worn on the garrison cap.

On 30 April 1945, Secretary of the Navy James Forrestall approved the publication "Uniform Regulations, U. S. Marine Corps Women's Reserve, 1945."

The regulations were a composite of the 1943 Regulations as modified by its changes together with several additional changes that appeared for the first time, and were to remain in effect until 1 October 1949.

A new item of clothing on which the ornaments were worn introduced by these regulations was the summer service L. S. (long sleeve) coat, prescribed for both officer and enlisted personnel.

Instructions contained in these regulations for the Officer's ornaments were:

CAP ORNAMENTS

(a) The service cap ornament shall be worn on the winter service cap, and on the summer dress cap with summer service and summer parade uniforms, and with the summer undress uniform when brown pumps are worn.

(b) .The dress cap ornament shall be worn on the summer dress cap with summer dress and summer dress A uniforms, and with the summer undress uniform when white pumps are worn.

(c) The left service collar ornament shall be worn on the garrison cap (and utility cap) on the left side, in the place provided therefor.

COLLAR ORNAMENTS

(a) Service collar ornaments shall be worn on the winter service coat centered on the collar• . .

(b) Collar ornaments shall be worn on the summer dress, summer service, and summer service L. S. coats centered on the collar• . • .

(c) Dress collar ornaments shall be worn on the summer dress coat, and on the summer service L. S. coat, when the latter is worn with white pumps.

(d) Service collar ornaments shall be worn on the summer servl, e coat, and on the summer service L. S. coat, when prescribed as summer parade uniform; also when the summer service L. S. coat is worn with brown pumps.

(e) Dress collar ornaments shall be worn on the summer dress A coat on the center line of each shoulder strap. . • .

For enlisted women the provisions concerning the usage of ornaments were:

CAP ORNAMENTS

(a) The bronze cap ornament shall be worn on the winter service cap, and on the summer dress cap with summer parade uniform.

(b) The gilt cap ornament shall be worn on the summer dress cap with summer dress and summer undress uniforms.

(c) The left bronze collar ornament shall be worn on the garrison cap (and utility cap) on the left side, in the place provided therefor.

COLLAR ORNAMENTS

(a) Bronze collar ornaments shall be worn on the winter service coat, centered on the collar. . .

(b) Collar ornaments shall be worn on the summer dress coat, summer service coat and summer service L.S. coat, centered on the collar• . . .

(c) Gilt collar ornaments shall be worn on the summer dress coat and on the summer L. S. coat.

(d) Bronze collar ornaments shall be worn on the summer service coat and on the summer service L. S. coat, when prescribed as summer parade uniform.

As in the case of the 1943 Regulations, the specifications for the prescribed ornaments were the same as for those worn by the male personnel of the period.

Figure 84. - Standard U. S. Marine Corps Emblem, approved 16 March 1936.

Figure 85. - Officer's dress cap ornament, 1937-1962.

Figure 86. - Officer's dress cap ornament, 1937-1962.

Figure 87. -Officer's dress cap ornament, 1937-1962.

Figure 88. -Officer's service cap and hat ornament, 1937-1942 ;
officer's service cap ornament, 1942-1962,

Figure 89. - Officer's service cap, and hat ornament, 1937-1942 , officer's service cap ornament, 1942-1962.

Figure 90. - Officer's service cap and hat ornament, 1937-1942; officer's service cap ornament, 1942-1962.

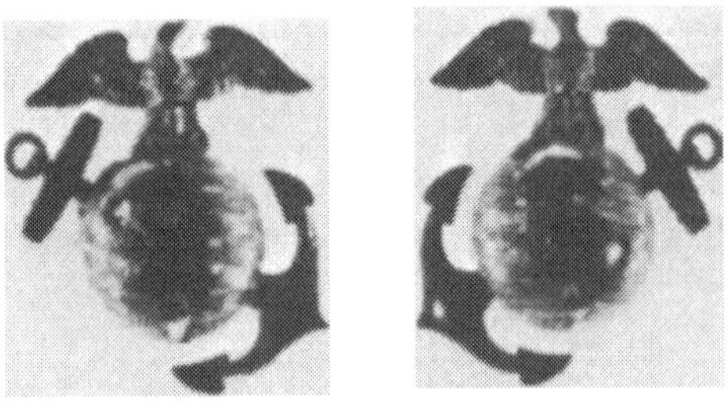

Figure 91. -Officer's dress collar ornaments, 1937-1962.

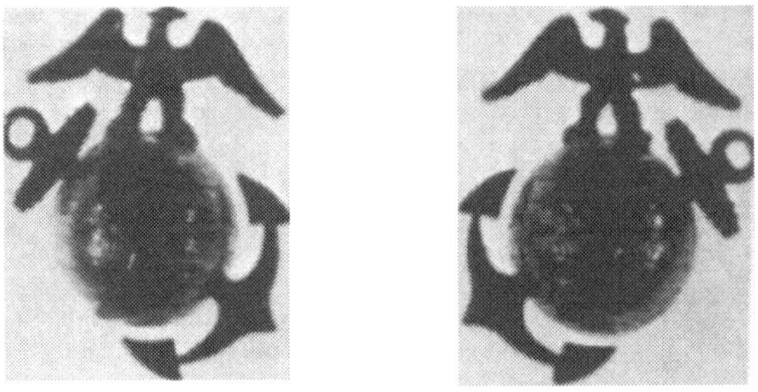

Figure 92. -Officer's dress collar ornaments, 1937-1962.

Figure 93. - Leader, Marine Band, 1937, in full dress uniform.

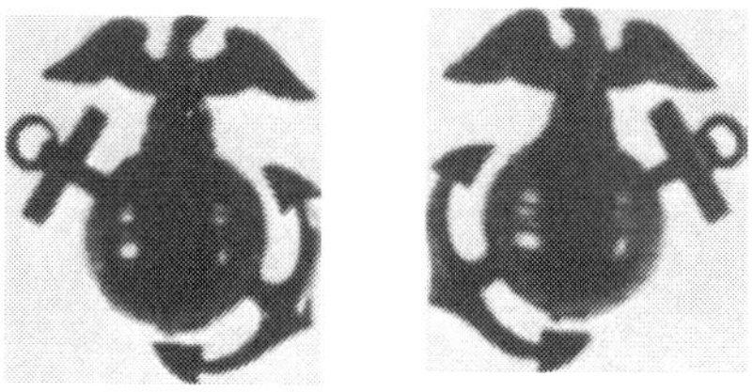

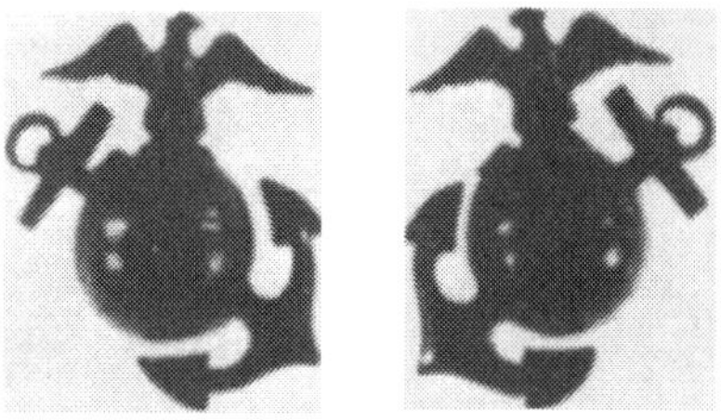

Figure 94. - Officer's service collar ornaments, 1937-1962.

Figure 95. - Officer's service collar ornaments, 1937-1962.

Figure 96. - Enlisted gilt cap ornament, 1937-1949; enlisted dress cap. ornament, 1949-1955

Figure 97. - Marine Corporal, circa. 1945, in blue undress "'B" uniform.

Figure 98. - Enlisted bronze cap and hat ornament, 1937-1942 ; enlisted bronze cap ornament, 1942-1955.

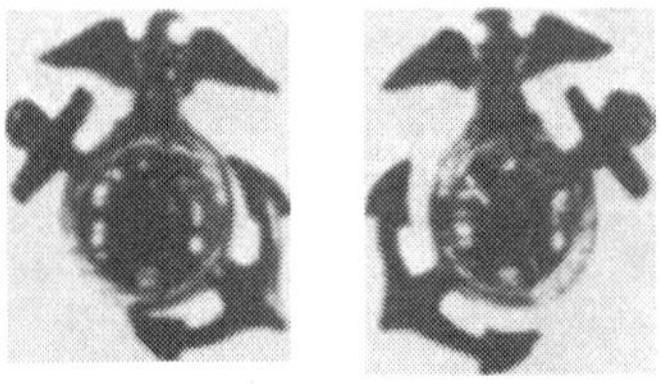

Figure 99. - Enlisted gilt collar ornaments (screw post type), 1937-1955.

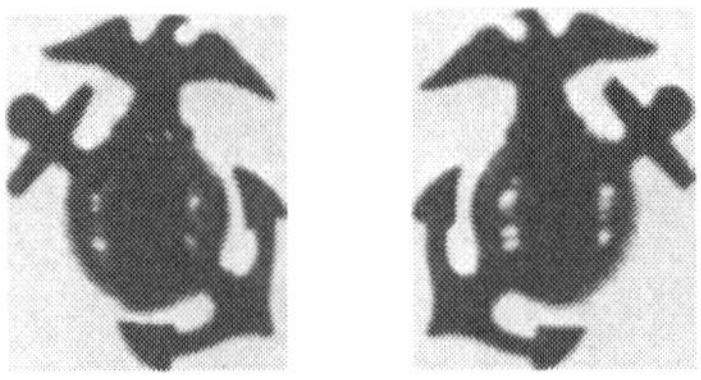

Figure 100. - Enlisted bronze collar ornaments (screw post type), 1937-1949; enlisted service collar ornaments (screw post type), 1949-1955.

Figure 101.-Marine Pfc, circa. 1945, in winter service "A" uniform.

Figure 102 . - 1\larine Corporal , circa . 1945, in swnmer service uniform with fiber helmet.

Figure 103. - Enlisted plastic service cap ornament, circa..1943.

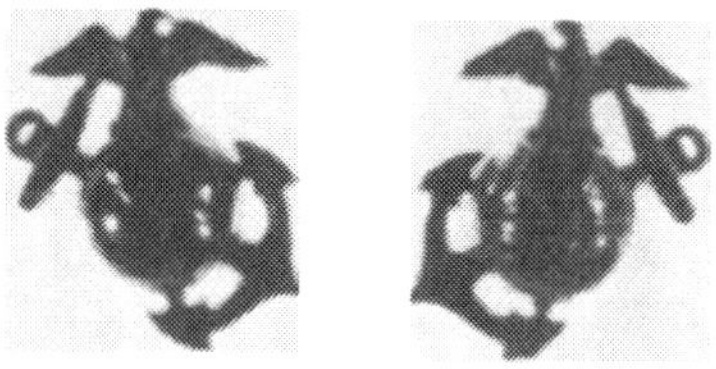

Figure 104. - Enlisted plastic service collar ornaments, circa. 1943.

Figure 105, - Service collar ornament (right) , (obverse) , circa. 1943, procured in Austr alia.

Figure 106. - S e r v i c e colla:r: ornament (right) , (reverse) , circa. 1943, procured in Australia, marked "K. O. LUKE MELBOURNE".

Figure 107. - Marine First Sergeant, 1946, in new winter servic:e "A" uniform with green wool service jacket and winter service garrispn cap.

Figure 108. - Marine Pfc, 1948, in summer service uniform.

Figure 109. - Marine First Sergeant, 1946, in blue undress "A,, uniform.

Figure 110. - Second Lieutenant Frances Seibert, USMCWR, 1944, in summer service "A" uniform.

Figure 111. - Cpl Beverly Jane Thybony, USMCWR, 1943, in winter service uniform.

Figure 112. - Sgt Mary Frances Igoe, USMCWR, 1944, in summer dress uniform.

Figure 113. - Woman Reserve Officer, 1944, in summer dress "A" uniform.

## CHAPTER IX

During the Post-World War II period, the status of uniform regulations was far from clear with two sets of regulations in existence, each with its own modifications or changes. The male Marines were using the 1937 Regulations as modified by Circular Letters, Letters of Instruction and a formal printed change, while at the same time the Women's Reserve was following the 1945 edition of its own regulations.

In order to consolidate all of these regulations into one publication, it was decided to include them in a chapter of the Marine Corps Manual which was then being revised. As a result, when the new manual was published in 1949, the new regulations appeared in Vol. I, Personnel and General Administration, as Chapter 49, "Uniforms".

Chapter 49 was a composite of the regulations of 1937 as modified, those of the Women's Reserve of 1945 with changes, plus a few additions appearing for the first time. The most notable of the pertinent new changes for male Marines was the abolition of the full dress uniform which was worn by officers detailed for duty at the White House. This uniform, on which the dress ornaments were worn on the full dress cap and epaulets, apparently fell into disuse during World War II.

For women officers, the summer dress, summer dress "A" and summer service L.S. (long sleeve) coats were redesignated as the short sleeve white, long sleeve white and long sleeve seersucker coats.

Instructions for wear of the cap and collar ornaments of male Marines as prescribed in the 1949 Regulations were:

### CAP ORNAMENT

1. The dress cap ornament shall be worn on the dress cap (with white or blue covers as well as the full dress cap of the Leader, Marine Band).

2. The bronze cap ornament shall be worn on the (winter and summer) service caps and the fiber helmet.

3. The left service collar ornament shall be worn on the left front side of the (summer and winter) service garrison caps, with the ornament centered vertically 2 inches from the front.

See Appendix M-1 for the cap ornaments' specifications.

### COLLAR ORNAMENT

1. Dress collar ornament shall be worn on the blue dress and white coats, the evening dress and white mess jackets (and on the shoulder knots of the Leader, Marine Band).

   a. On the blue coat, the white coat, and the evening dress jacket, the ornaments shall be placed vertically in the center of each side of the collar• . .

b. On the white mess jacket, the ornaments shall be worn on the center line of the shoulder strap• • .

2. Gilt collar ornamer ts shall be worn by all enlisted men on the dress coat and by members of the Marine Band on the full-dress coat, except the full dress coat worn by the Drum Major.

3. Service collar ornaments shall be worn on the summer and winter ser- vice coats and jackets.

4. Collar ornaments will be not worn on khaki shirts.

The specifications for the collar ornaments are contained in Appendix M-2.

For the women officer's ornaments the regulations provided:

CAP ORNAMENT

1. The bronze cap ornament shall be worn on the winter service cap, and on the summer dress cap with summer service and with the summer undress uniforms when brown pumps are worn.

2. The dress cap ornament shall be worn on the summer dress cap with summer dress and summer undress unif orms when white pumps are worn.

3. The lef t service collar ornament shall be worn on the lef t front side of the garrison cap (and utility cap) in the place provided therefor.

COLLAR ORNAMENT

1. Service collar ornaments shall be worn on the winter service coat, cen- tered on the collar• • •

2. Collar ornaments shall be worn on the summer dress "A", summer undress "A", and "B", and summer service coats, centered on the collar• . •

3. Dress collar ornaments shall be worn on the summer dress "A", summer undress "A", and summer undress "B" when worn with white pumps.

4. Service collar ornaments shall be worn on the summer service and summer undress "B" coats when worn with brown pumps.

5. Dress collar ornaments shall be worn on the summer dress "B" and summer undress "C" coats on the center line of each shoulder strap. • , .

The manner of wearing the enlisted women's ornaments as provided for in the regulations were:

CAP ORNAMENTS

1. The bronze cap ornament shall be worn on the winter service cap, and on the summer dress cap with summer service or summer undress "B" when worn with brown pumps.

2, The gilt cap ornament shall be worn on the summer dress cap with summer dress and summer undress uniforms when white pumps are worn.

3, The lef t bronze collar ornament shall be worn on the lef t front side of the garrison cap (and utility cap) in the place provided therefor,

COL LAR ORN AJ\IENT

**1.** Bronze collar ornaments shall be worn on the winter service coat, centered on the collar, , .

2, Collar ornaments shall be worn on the summer dress, summer undress, and summer service uniforms, centered on the collar, , ,

3, Gilt collar ornaments shall be worn on the summer dress coat, summer undress "A", and summer undress **"B"**, when whi te pumps are worn.

4. Bronze collar ornaments shall be worn on the summer service coat, and on summer undress **"B"**, when brown pumps are worn,

The ornaments initially authorized for wear by these regulations were the same as those presc ribed by those of 1937 (see Chapter VIII),

During the period the 1949 Regulations were in force (1949-1962) important changes took pl ace involving the ornaments as well as the items on which they were worn,

Du ring the later part of 1951 or the early part of 1952 the enlisted khaki service jacket was phased out, In its pl ace, a tropical garrison khaki shirt was authorized , which had a clutch type service collar ornament, These ornaments were similar in design to those prescribed in 1937 but incorpor ated clutch fasteners. (Figs. 114 & 115) Shortly thereaf ter the clutch service collar ornament was presc ribed for the khaki (cotton) shirt and in addition, under certain circumstances , the clutch type gilt collar ornaments could be worn on the tropical or khaki shirts when worn with the blue trousers. (Fig. 116) Printed Change No. 4 of January 1954 reflec ted the addition of the collar ornaments to the shirt collar s of enlisted men by amending paragr aph 4 of the ORAMEK T, COLLAR section to read:

4, Collar ornaments will not be worn on khaki shirts, except when Summer Service "A" and **"B"** or Winter Service "C" is presc ribed at which time enlisted men will wear the service collar ornaments centered on each side of the shirt collar midway between top edge of collar and point, and one (1) inch from edge of collar , with wings and eagle par allel to the top edge of the collar,

In the 1952-53 period a new set of uniforms, the M-52 series, was presc ribed for Women Marines. These were worn concurrently with the "old style" uniforms, which were to phase out by mid 1957. At the time the new series was introduced , some of the articles of clothing under the old system were redesignated or discon- tinued, These included the summer dress cap that was renamed the summer service cap, the winter service coat that was redesignated as the green coat, and the summer service coat that was renamed the green and white stripe coat. In addition, tne short sleeve white coat was redesignated the white coat, while the long sleeve white coat was phased out. Also at this time the long sleeve seersucker coat disappeared, along with that of the short sleeve seersucker coat for enlisted women. Under the M-52 series the Women Marines "dress blue" uniform appeared with the appropriate dress or gilt ornaments being worn on the blue cap and coat. (Fig. 117).

An evening dress uniform for women officer' s was also prescribed by these regulations, together with the requirement that the dress collar ornaments be worn on the evening dress jacket.

Other uniforms prescribed were the new winter service with its green cap and coat with appropriate service ornaments; (Fig. 118) and the summer service uniform with its dacron-nylon cap, coat and dress. The service collar ornaments were worn on both the coat and dress, while the left service collar ornament was worn on the cap. (Fig. 119)

Due to the various modifications in the Women Marines uniforms, the paragraphs pertaining to their ornaments were amended by Printed Change No. 5 of August 1954. These revisions are shown in Appendix M-8.

Printed Change No. 5 also officially recognized the practice of wearing the cap ornament on the helmet liner. Plate No. 43 contained an illustration of an enlisted. member of the Military Police wearing a white enameled helmet liner encircled with a scarlet and gold band, with the enlisted gilt cap ornament placed in front. This helmet liner was authorized only for "Fleet Marine Force units in the field at the discretion of the Commanding General", in accordance with the Commandant's remarks to the Uniform Board's Minutes No. 4 of 26 July 1954.

Several additional changes appeared in Printed Change No. 6 of December 1954 involving articles of uniform on which the ornament was worn. The most notable of the changes was the abolition of the officer's khaki service jacket. Another was the modification of the officer's mess jacket with the dress collar ornaments being removed from the shoulder straps and placed on the lapels. Also, white cap covers as well as blue were prescribed for enlisted members of the Marine Band in special full-dress uniform. Further, the service collar ornaments were removed from the khaki shirt as part of winter service "C" uniform and the regulations were modified as follows:

> "Collar ornaments will not be worn on khaki shirts, except when Summer Service "A" and "B" is prescribed, at which time enlisted men will wear the service collar ornaments centered on each side of the shirt collar• • •"

Printed Change No. 9 of August 1955 amended the section of the regulations pertaining to the wearing of collar ornaments to reflect the change in their placement on their white mess jacket which had previously been authorized in Printed Change No. 6 of December 1954. The wording of the pertinent sub-paragraph was revised to:

> "On the white mess jacket, the ornaments shall be worn on the collar, in the eyelets provided therefor• • •"

In Printed Change Number 14 of June 1956, the term "Ornament" was changed to "Insignia, Branch of Service" and all appropriate paragraphs in the regulations were modified accordingly. Although new terminology was used there were no essential differences in the revised paragr aphs concerning the ornaments or manner in which they were worn. See Appendix M-11 for a description of the Insignia, Branch of Service by this Change.

By 30 June 1957 most of the old-style women Marines uniforms had phased out eliminating several items of uniforms on which the ornaments were worn. These included the old-style winter service cap and green coat, as well as the green and white stripe coat. The enlisted summer service cap and white coat also had been abolished by that time.

Several new changes involving items of uniform on which the ornament was worn were contained in Marine Corps Order 1020.24 of 29 August 1957. The most significant change discarded the blue cap cover 1October 1957, for all male personnel except those of the Marine Band. Other changes authorized green and khaki coats with bellows backs for officers, and tropical and khaki shirts with french cuffs for staff non-com issioned officers.

Several changes involving uniform items were to take place in 1958. MCO 10123.5 of 15 January 1958 prescribed a redesigned white uniform for women officers. This uniform included a new style long sleeve white coat on which the dress collar ornaments were to be worn.

In July 1958, Printed Change No. 25 of Chapter 49 was issued. This change specified that the "jacket, man's green" (winter service jacket) worn since 1946 would be obsolete by 30 June 1960. However , due to exigencies of the period, this jacket was authorized to be worn under certain circumstances until 1968.

In 1958, a green cap ofthe same style and design as the winter service and blue . dress caps was approved for wear by women officers with their white uniform. Information concerning this cap, on which the dress cap ornament was worn, is contained in the Uniform Board's Minutes to Meeting No. 33 held on 23 July 1959.

During the period 1959-60 , additional changes or modifications occurred involving pertinent uniform items. One significant change was contained in MCO 1020.27 of 26 February 1959 which introduced "a 14 ounce blue gabardine material as standard for the manuf acture of the male enlisted blue uniform." Thus, the use of kersey material which h ad been used in the manufacture of the enlisted blue uniform since before the turn of the century had come to an end.

Also during this period a new item of uniform was produced on which the ornament was worn. MCO 10120.23A of 9 July 1959 and MCO 1020.29 of 13 April 1960 introduced the short sleeve shirt for male Marines into the supply system. The method of wearing the ornament on the enlisted short sleeve shirt (shirt, khaki, with quarter length sleeve) is illustr ated in Figure 120 which was taken from MCO 10120.23A.

Additional changes concerning articles of clothing on which the ornament was worn appeared in Printed Change No. 44 of February 1961. The field hat which had been abolished on 28 May 1942, reappeared bearing the new identificati,. •"Hat, Service, Campaign." This hat, on which the service cap ornament was worn, was authorized for wear only by rifle and pistol team members, permanent range personnel and Drill Instructors at the Recruit Depots at Parris Island and San Diego. The paragraphs dealing with the ornaments (Insignia, Branch of Service) worn by male Marines were also revised. While no major changes were encountered, the collar ornaments could again be worn on the shirt collars with the enlisted winter service "C" uniform. See Appendix M-12 for these revisions•

. Another important change was contained in Marine Corps Order 1020.33 of 5 May 1961, which removed the ornaments from the shirt collars of enlisted men. This resulted from Commandant General David M. Shoup's approval of the recommendations of a Staff Non-Commissioned Officers Symposium presented to him by a memorandum of 11 January 1961 enclosing the Minutes of the Permanent Marine Corps Unifor m Board Meeting No. 43.

In 1961, the uniform regulations again became a separate publication. Marine Corps Order P1020.34 of 2 June of that year , removed Chapter 49 in its entirety from Vol. I of the Marine Corps M anual and redesignated it the "Interim Uniform Regulations." The order further stated that: "Headquarters is currently preparing a new manual entitled ' Marine Corps Uniform Regulations' to replace the 'Interim Uniform Regulations."' Those were to appear in 1962.

During the tenure of the Interim Uniform Regulations one major development occurred involving the ornaments.

This was contained in the Uniform Board's Minutes to Meeting Ne,. 49 held on 20 July 1961 in which the Quartermaster General of the Marine Corps p.:oposed to "designate the Insignia, Branch of Service, collar clutch as limited standard and adopt the Insignia, Branch of Service, collar screw post as the standard Insignia. "Background for this proposal as contained in the minutes:

1. On 23 January 1961, the Commandant of the Marine Corps approved the Uniform Board's recommendation to discontinue the wearing of collar ornaments on the khaki and tropical shirts.

2. In view of the above, the Quartermaster General has recommended:

a. That the Insignia, Branch of Service, collar clutch be declared limited standard and that the Insignia, Branch of Service, collar, screw post be adopted as the standard insignia.

b. That Insignia, Branch of Service, collar clutch, gold be declared obsolete and disposed of since all requirements for this type insignia have been deleted.

3. The clutch· type insigni a as well as the screw post type is presently being utilized on enlisted green coats, jackets and garrison caps. The screw post type insignia only is being utilized on blue uniform collars. It is not considered feasible to utilize the clutch type insignia with the blue uniform collar.

4. The Military Clothing and Textile Supply Agency is in a buy position for Insignia, Branch of Service and unless action is initiated to change supply classifications, the clutch type insignia will be procured.

After considering this background information, the Board arrived at the following recommendations:

a. That the Insignia, Branch of Service, Bronze, collar clutch be designated as limited standard and Insignia, Branch of Service, Bronze, collar screw post be adopted as the standard insignia.

b. That the Insignia, Branch of Service, Gold, collar clutch be declared obsolete and disposed of to the best interests of the Government,

These were then submitted to the Commandant on 28 August 1961 who in turn approved them on 6 October of that year.

## SEAL OF THE UNITED STATES MARINE CORPS

In 1954 a chain of events took place that led up to the last major change in the design of the ornament. On 22 June of that year President Dwight D. Eisenhower signed Executive Order 10538 "Establishing a Seal for the United States Marine Corps." Background behind the origin of this seal was contained in a press release of that time:

"President Eisenhower today approved the design of an official seal for the United States Marine Corps.

The adoption of this seal marks the first time in 178 years of Marine Corps history that the Corps has had one standard official seal.

The new seal consists of the traditional United States Marine Corps emblem in bronze; an American bald eagle with wings displayed standing upon the western hemisphere of the terrestrial globe holding in his beak a scroll inscribed with the Marine Corps motto "SEMPER FIDELIS" (Ever Faithful) with the hemisphere superimposed on a foul anchor. The emblem is displayed on a scarlet background encircled with a Navy blue band edged in a gold rope rim and inscribed Department of the Navy, United States Marine Corps in gold letters.

The seal was designed at the request of the Commandant, General Lemuel C. Shepherd, Jr., by the Heraldic Branch of the Quartermaster General, Department of the Army, who is charged with all heraldic matters for the armed forces by the Department of Defense. The design of the seal incorporated minor and technical refinements based on the STANDARD U. S. MARINE CORPS EMBLEM adopted in 1936," (The seal is illustrated in Fig. 121.)

Coincident with the approval of the seal by the President, the Marine Corps Emblem centered on the seal was adopted as the official Marine Corps Emblem. As a result, the Commandant in 1955 approved the enlisted men's cap and collar ornaments developed by the Permanent Marine Corps Uniform Board which incorporated the new design. The necessary dies and tools to manufacture these ornaments were then procu red and transmitted to the Marine Corps Supply Activity where they could be loaned as needed to Contractors for the manufacture of the enlisted ornaments.

Shortly thereaf ter; the new enlisted cap and collar ornaments began to appear in the supply system. The anchor of these ornaments contained reversed flukes, which made it technically correct for the first time since 1868. Previous orna- ments had both anchor flukes turned outboard.

The new collar ornaments included a rope wrapped around their anchors, which was the first time that anchor ropes appeared on collar ornaments.

The color of the ornaments for the dress uniform was gilt or gold and similar in color to those previously issued. The service ornaments were a dark-brown color instead of the previously issued black ornaments.

The gilt and service cap ornaments contained a screw post fastener. However , the service collar ornament contained a clutch fastener while those of gilt .were issued with both clutch and screw post fasteners. ( Fig. 122) The gilt ornament with the clutch fasteners was worn on the shirt, while the one with the screw post fastener was worn on the blue coat. (See Figures 123, 124, 125, 126 & 127)

Lack of uniformity in officer's ornaments led to a study by the Uniform Board of the feasiability of a standard ornament. Previously each insignia manufacturer , although keeping within specified dimensions, produced an ornament that diff ered in design from all others. Af ter a detailed study, the Board submitted a memoran- dum to the Corrunandnat on 5 May 1960 recommending that the officer's ornaments submitted with the memo, and based upon the design of the Marine Corps E mblem in the M arine Corps Seal, be approved as standard, which will enable uniformity to be "achieved throughout the Marine Corps and industry," The board's recom- mendations were approved by the Commandant, General David M. Shoup on 10 May 1960.

Due to necessary retooling operations on the part of the manufacturers, it wasn't until 1962 that thE! new officers ornaments appeared. As with the new en- listed ornament, the anchor flukes on these were reversed, and the off icers collar ornaments contained anchor ropes for the first time. See Figures 128, 129, 130 & 131 for specimens of these ornaments.

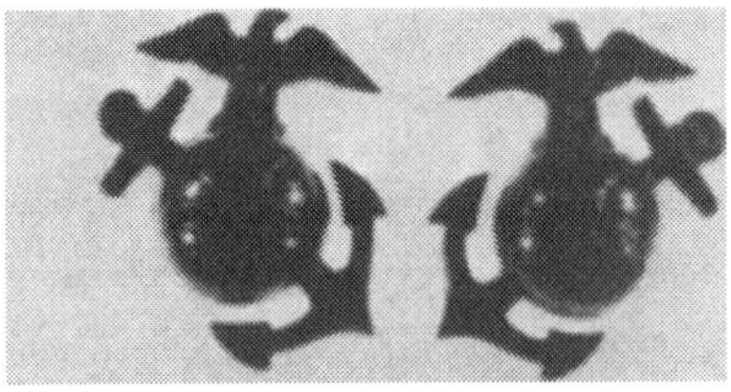

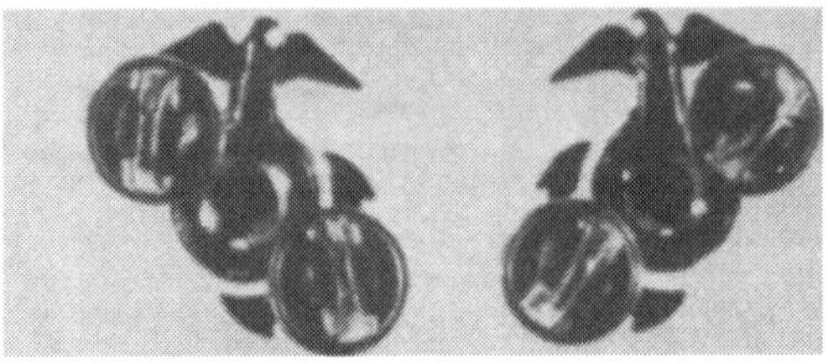

Figure 114. Enlisted service collar ornaments (clutch) (obverse), 1951-1955.

Figure 115. - Enlisted service collar ornaments (clutch) (reverse), 1951-1955.

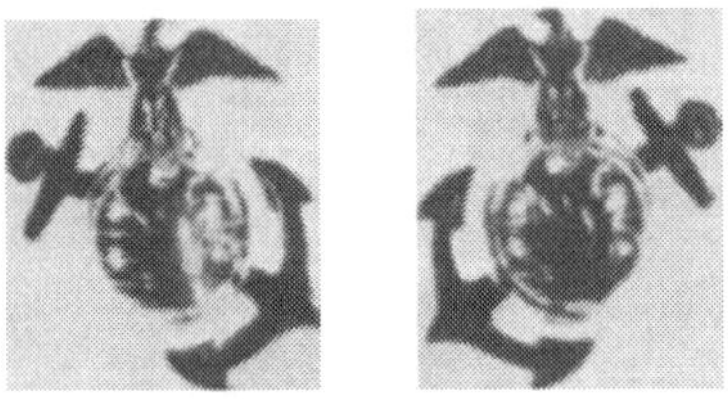

Figure 116. - Enlisted gilt collar ornaments (clutch) circa. 1951-1955.

Figure 117. - Woman Marine Officer, 1952, in blue undress uniform.

Figure 118. - Woman Marine Captain 1952, in winter servwe "A" uniform.

Figure 119. - Women Marines' Uniforms, 1952. The Woman Marine, left, models the blue undress uniform. In the center is the winter service "A" uniform, while on the right is the summer service uniform.

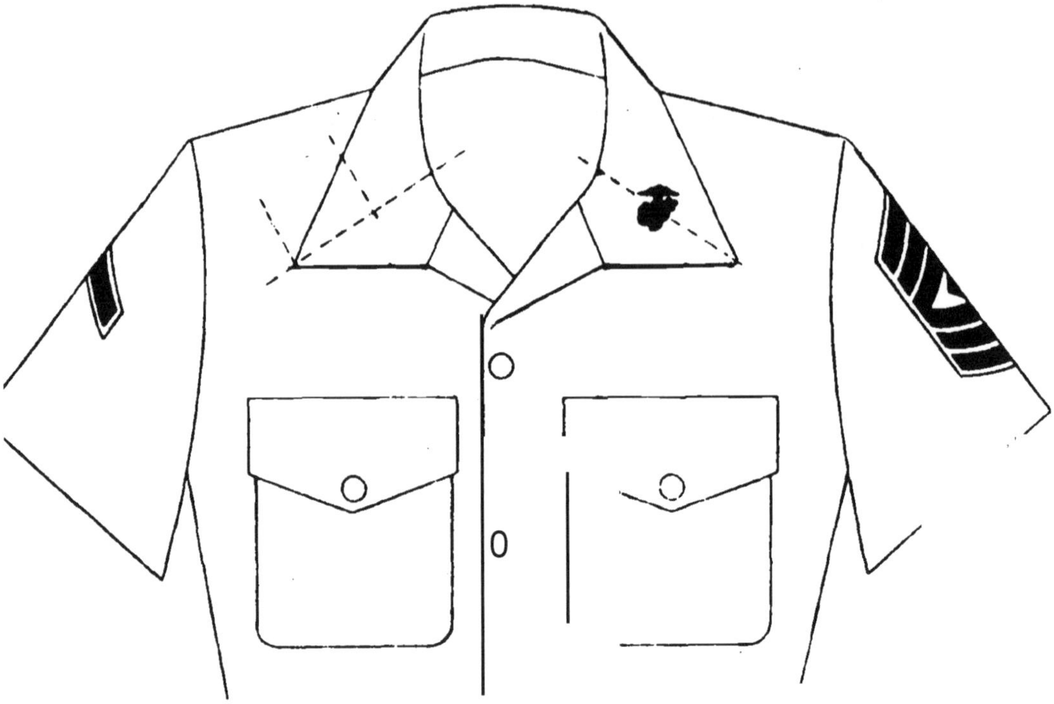

Figure 120. - Correct placement of enlisted collar ornaments on the short-sleeve, khaki shirt.

Figure 121. - Seal of the United States Marine Corps, approved 22 June 1954 by President Dwight D. Eisenhower.

Figure 122. - , Marine Corporal 1960, in summer service "A" uniform.

Figure 123. - Enlisted dress cap ornament, 1955-1956; enlisted dress cap insignia, branch of service, 1956 to the present.

Figure 124. - Enlisted bronze cap ornament, 1955-1956; enlisted bronze cap insignia, branch of service, 1956-1961; enlisted service cap insignia, branch of service, 1951-1963,

Figure 125. - Enlisted gilt collar ornaments (clutch) , 1955-1956; enlisted gold collar insignia, branch of service (clutch) 1956-1961; enlisted dress collar insigni a, branch of service (clutch) , 1961-1961.

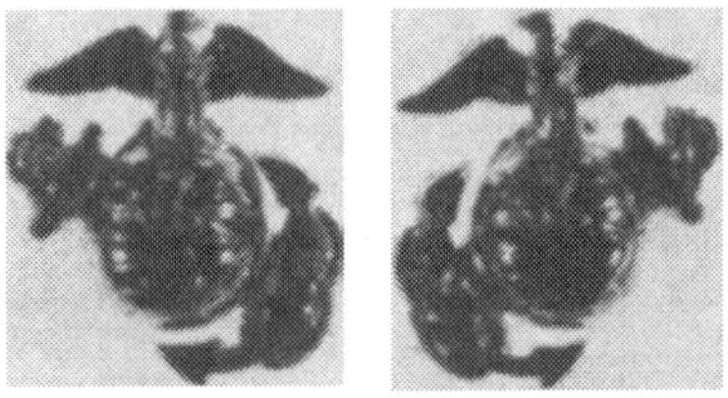

Fignre 126. - Enlisted gilt collar ornamc.1s (screw post) , 1955-1956; enlisted gold coll ar insignia, branch of service (screw post) , 1956-1961; enlisted dress collar insignia, branch of service, 1961 to the present.

Figure 127.– Enlisted service collar ornaments (clutch) , 1955-1956; enlisted bronze . collar insignia, branch of service (clutch) , 1956-1961; enlisted service collar insignia branch of service (clutch) , 1961-19 63.

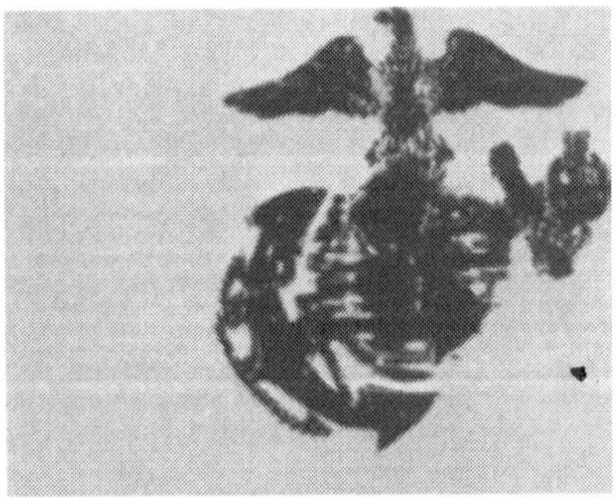

Figure 128. - Officer's dress cap insignia, branch of service, 1962 to the present.

Figure 129. - Officer's service cap insignia, branch of service, 1962-1963.

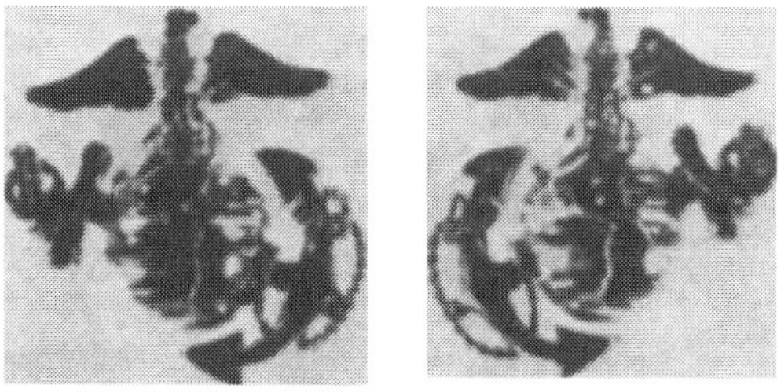

Figure 130. - Officer's dress collar insignia, branch of service 1962 to the present.

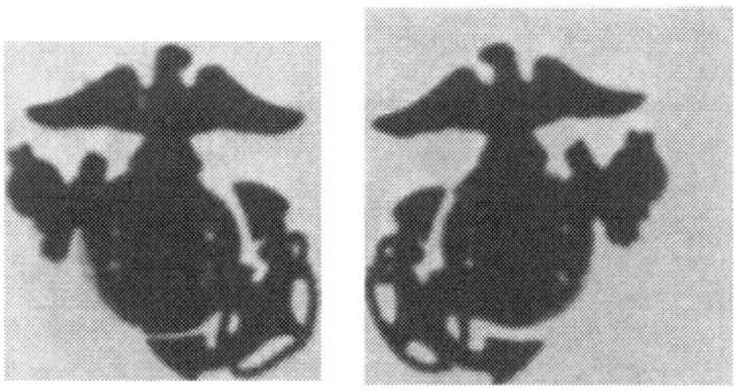

Figure 131. - Officer's bronze collar insignia, branch of service. 1'962-1963.

## CHAPTER X

A little over a year after being separated from the Marine Corps Manual, the 1961 Interim Uniform Regulations were replaced by a publication entitled "Marine Corps 'C"nifor m Regulations." This was contained in a binder and was transmitted to the field by Marine Corps Order P1020.34A of 2 October 1962.

Although the new regulations were in an entirely new format, they contained information that was similar to those of 1961 plus a few minor modifications. These included changes to or additions of articles of uniform on which the ornament was worn.

Among the changes for male Marines was the elimination from the regulations for the Officer's service jacket as well as the requirement for wearing the ornament on the enlisted shirt collar. This requirement had previously been terminated by MCO 1020.33 of 5 May 1961.

In addition to these changes, there were also new regulations for the Marine These Band. included the authorization of full dress coats in both summer and weights as winter well as the requirement for the first time for (dress) collar ornaments to be worn on the full-dress coat of the Drum Major. Also, a scarlet tropical worsted full-dress coat was prescribed for wear by members of the Drum and Bugle Corps of the Marine Corps (stationed at the Marine Barracks, Washington, D. C.). This organization had previously worn the enlisted dress blue coat. (Fig. **132)**

The changes for Women Marines included the redesignation of the summer service cap worn by officers as the dress green cap. Also, a new uniform, the Summer Service Optional, was prescribed for women officers. This was similar in description to the Summer Service uniform, except that the Dacron-Nylon Coat was required to be worn, and the dress green cap with bronze buttons and service insignia was substituted for the Dacron-Nylon cap.

The sections of the regulations pertaining to the ornaments worn by male Marines were similar in scope to those of their predecessor, although they were arranged in an entirely different manner. See Appendices N-1 & N-3.

For Women Marines the instructions for the use of the ornaments were completely revised. This resulted from the fact that when the old-style uniforms were phased-out, the sections of the previous regulations were not revised at that time to reflect the change. Also the new instructions picked up the change in terminology of the officers summer service cap to that of the dress green cap. These are contained in Appendices N-2 & **N-4.**

Following the promulgation of the new Regulations, changes were issued in the form of Printed Changes, Marine Corps Orders and Marine Corps Bulletins that modified both the ornaments and the iterns of uniform on which they were worn.

Also during 1962 a change eliminated the white military police helmet as an item for wear with the service cap frame and white cap cover being substituted in its place. This action was brought about by a recommendation contained in

Minutes of the Permanent Marine Corps Unifor m Board Meeting No. 55 of 11 October promulgated to the field by Marine Corps Bulletin 1020 of 7 June 1963. In order to reflect this change , the section of the regulations pertaining to enlisted ornaments was modified by Change 2 of 7 }.::bruary 1964 as follows:

2. The dress cap insignia shall be worn on the frame, service cap when the white cap cover is worn.

The last modification of the ornament to bring it to its present form began with the publication of MCO 1020.37 of 4 June 19 63, which changed the color of the service ornament from brown to black. The order pointed out that in line with a Depar tment of Defense policy of achieving maximum economy th rough standard usage wherever possible, a Marine Corps study found that in the area of uniform clothing:

" • .significant economy can be ·effected through use of black uniform accessories, without loss of utility and with a minimum change in appear ance."

The order then provided for the color transition from brown to black of uni- form accessories such as shoes, socks, service cap frames, gloves and the womens handbags.

With ref erence to buttons and ornaments , the order stated that a study was being conducted to see which of those items needed to be changed as well as the most economical phase-in methods. The results of this study were contained in Marine Corps Bulletin 1020 of 12 December 19 63 which provided that:

Non-leather items shall remain the same until existing stocks are exhausted. Further procurement will be in non-glossy black. A cut-off date has not been established for these items. The only restriction is that when an item is re- placed in black color on a uniform, all like items shall be replaced at the same time, i.e., when one button is replaced in black color , the entire set of buttons will be replaced.

As a result, black service ornaments started appearing in both the supply system and in the Marine Exchanges during the summer of 1964. (Figs. 133, 134, 135 & 136)

Another important change that took place in 1963 was in the material used in the manuf acture of the winter service uniform worn by enlisted men. Marine Corps Order 10120.35 of 1 July 1963 authorized that the issue and sales of a new 15 ounce green wool serge uniform would begin on 1 July 1963 where the supply of the kersey unifor m was exhausted. The order further authorized this new uniform, subject to minor modifications, for wear by officers. It also stated that during the transi- tion period of changing from uniforms of kersey to uniforms of serge, that **mixing** of garments of the two materials was prohibited. Finally the order specified that phase-out of the kersey uniform was to be completed by 1 July 1967. Thus the use of the green kersey material that had been used in the manufacture of the enlisted mans winter service uniform since 1914 had come to an end.

One last change that occurred in 1963 concerned the male officer's bellows back coat. Marine Corps Bulletin 1020 of 10 September 1963 specified that effective immediately that coat would be no longer authorized. However officers who possessed that coat prior to the publication of the bulletin could continue to wear it until it became unserviceable.

At Uniform Board Meeting No. 64 of 26 May 1964, two new women officers uniforms were recommended for wear. These were a redesigned evening dress uniform and a newly introduced mess dress uniform, on the jackets of which were to be worn the dress collar ornaments. These recommendations were submitted to the Commandant on 12 June 1964 who approved them four days later.

A new change concerning the wearing of ornaments took place with the publication of Marine Corps Bulletin 1020 of 6 July, 1965. This bulletin authorized Navy officers attached to Marine Corps organizations to wear the service uniforms prescribed for officers of the Marine Corps. When doing so, the Marine Corps officer's service collar ornaments would be worn on the service coats, while the bronzed Navy miniature cap device would be worn on the service and garrison caps. This practice, subject to minor modifications, was applied during both the WWI and WWII eras.

On 22 April 1966 a new set of uniform regulations was announced by Marine Corps Order P1020.34B. These regulations were almost identical in format and content to those of 1962 as modified by all printed changes and directives. New changes did appear for the first time in these regulations that involved items of uniform on which the uniform was worn.

Among the new changes that appeared was the notification that the green jacket worn by enlisted men would be obsolete and no longer worn after 1 July 1968. Also contained in the regulations was the authorization for Navy officers to wear Marine Corps service uniforms with Corps service collar ornaments to be worn on the service coats. This was previously prescribed by Marine Corps Bulletin 1020 of 6 July 1965.

The new mess dress uniform for women officers, along with a newly designed evening dress uniform, also appeared in these regulations. These had previously been approved by the Commandant on 16 June 1964.

As a result of these changes the sections of the regulations pertaining to the ornament were modified accordingly and while no major changes were encountered some of the terminology was changed to a very slight degree. See Appendix 0.

Since the introduction of the 1966 Uniform Regulations several changes have taken place involving items of uniform on which the ornament was worn. Two of these changes appeared in Change 26 to Marine Corps Order P10120.28 (Individual Clothing Regulations) of 19 June 1967 that authorized new summer service and utility uniforms for the Women Marines.

The new summer service uniform was manufactured of a green and white polyester/cotton cloth and consisted of a skirt and short sleeve jacket on which the service collar ornaments were worn. The issuing of this uniform began on 1 July 1967, and was to completely replace the old-style dress and long sleeve coat during Fiscal Year 1969. This new uniform had previously been approved by the Commandant on 2 May 1965, after reviewing the recommendations of the Uniform Board's Meeting No. 68 of 25 March 1965.

The new utility uniform for Women Marines consisted of a light blue shirt, dark blue slacks and a dark blue garrison cap on which the left service collar ornament was worn. Issuing of this uniform began on 1 July 1967, and it was to replace the green utility uniform by 1 July 1971. The garrison cap for this uniform was recommended by the Uniform Board in the minutes to their Meeting No. 60 held on 17 September 1963, and was approved by the Commandant, General David M, Shoup on 14 October of that year.

Two additional changes to the 1966 Uniform Regulations involving the wearing of the ornament appeared in Chang 2 of 9 April 1968, One change eliminated the khaki cap cover with dress ornament from the cap frames of Honor Guard personnel and substituted the white cap cover with dress ornament in their place, The other reflected the substitution of the blue garrison cap for the green utility cap worn by Women Marines, which resulted in the section of the regulation pertaining to the wearing of the ornaments by enlisted women being modified to read:

> 6, The left service collar insignia shall be worn on the left front side of the garrison caps in the eyelet provided,

## SUMMARY

As the symbol of the remarkable esprit of the U. S. Marine Corps the Eagle, Globe and Anchor has survived its first century with honor and acclaim and consequently it has been chosen as the basis for the emblem of amphibious forces of nations throughout the free world. As shown in the text of the study the change in the emblem since its introduction in 1868 have been largely minor in nature never really departing from its initial design.

Figure 132. - Musician, U.S. Marine Band, 1965, in full-dress uniform.

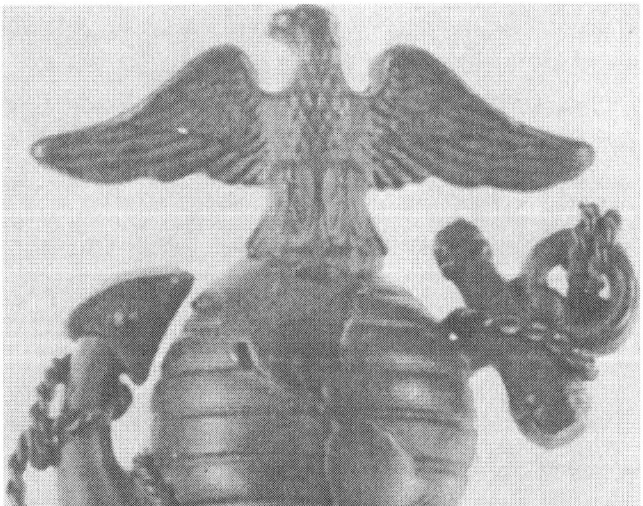

Figure 133. - Officer's cap insignia, br anch of service, 1963 to the present.

Figure 134. -Officer's service collar insignia, branch of service 1963 to the present.

Figure 135. - Enlisted service cap insignia, branch of service, 1963 to the present.

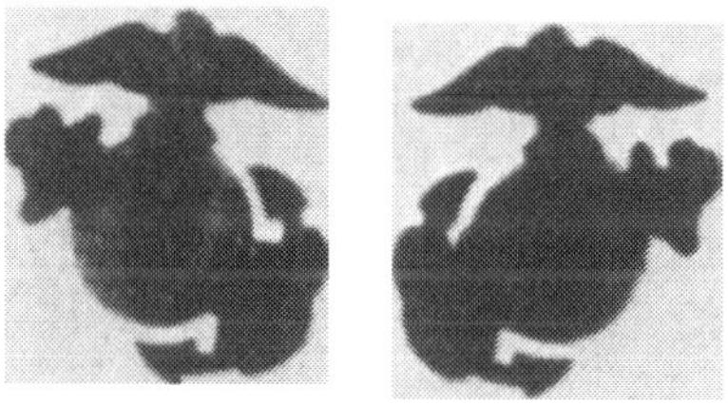

Figure 136. - Enlisted service collar insignia, branch of service, 1963 to the present.

## APPENDIX A

## THE 1868 FATIGUE CAP ORNAMENT

(Extract from a letter from Maj Augustus S. Nicholson, Adjutant and Inspector, HQMC, to eleven Corps officers commanding posts and stations, dated 30 Nov. 1868,)

### DESCRIPTION OF NEW FATIGUE CAP ORNAMENT

For Commissioned Officers: - Afretted silver hemisphere struck from solid plate with chased parallels and Continent of North and South America of gold plate; to be surmounted by a spread eagle cut from solid silver and securely fastened by means of a tang of silver soldered on the inner side of the hemi- sphere. The hemisphere to rest upon the shank, between the stock and flukes of a foul anchor worked in gold bullion. The dimensions to be as follows: hemisphere three fourths (3/4) of an inch in diameter, the eagle three eighths (3/8) of an inch from top of the head to point of claws. Spread of wings one (1) inch, Shank of anchor to intersect the hemisphere between the Arctic Circle and the Tropic of Cancer on the left, and the Equator and the Tropic of Capricorn on the right. Anchor over all in length one and one half (1 1/2) inches, one eighth (1/8) of an inch in width, the inner edge of flukes and stock touching the hemisphere. The arms of anchor including flukes to be eleven sixteenths (11/16) of an inch in length from point of fluke to point of shank and one eighth (1/8) of an inch in width, The whole to be on a ground work of dark blue cloth,

For enlisted men, same as for Officers except that the whole be struck from plate brass 1/16 inch thick,

## APPENDIX B

## THE 1869 OFFICER'S EPAULETTE ORNAMENT

Headquarters Marine Corps,
Adjutant and Inspector's Office,
Washington, Nov. 20, 1869.

Ordered:

By direction of the Navy Department, there will be substituted for the Silver Bugle worn by officers of the Marine Corps on the Epaulette, an ornament of the same design as that at present worn by the officers on the fatigue cap, except that the anchor be of plain silver, and the ornament to be of the following dimensions:

| | |
|---|---|
| Length of anchor, over all | 2 inches |
| Length of anchor stock | 1/2 inch |
| Diameter of globe | 9/10ths of an inch |
| Height of eagle | 4/10ths of an inch |
| Spread of wings, from tip to tip | 1 inch |

Deviations from regulations having been observed in the cap ornament, as also in the size, shape, and style of gold cord of the shoulder knots worn by many officers, attention is called to the importance of uniformity; and Commanding Officers will require strict compliance with the patterns and designs prescribed for the Corps.

By order of the Brig. General Commandant:

AUG. S. NICHOLSON,
Major U. S. M. C.
Adjutant and Inspector.

## APPENDIX C

## THE 1875 UNIFORM ORNAMENT

(Extracts from Regulations for the Uniform and Dress of the Marine Corps of The United States, May, 1875.)

1. Officer's Full Dress Hat Ornament

   Ornament on front of hat -- a richly-gilt United States shield, two and seven-eighths inches long by two and three eighths inches wide; resting upon the centre of the shield a silver corps device of same design and dimensions as that prescribed for the officer's undress cap.

2. Officer's Undress Cap and Fatigue Cap Ornaments

   CAP ORNAMENT

   A fretted silver western Hemisphere, struck from solid plate, with chased parallels; continents of North and South America of gold plate. To be surmounted by a spread eagle cut from solid silver, and securely fastened by a tag silver soldered on the inner side of the hemisphere. The hemisphere to rest upon the shank, between the stock and flukes of a foul anchor embroidered in gold bullion; the inner edges of the flukes touching the hemisphere. The dimensions to be as follows: hemi- sphere, three-fourths of an inch in diameter; the eagle, three-eighths of an inch from top of head to point of claws; spread of wings, one inch; width of anchor stock and shank, one-eighth of an inch; arms of anchor, including flukes, three-fourths of an inch Jn length from the peak of flukes to apex of crown, and one-eighth of an inch in width; length of anchor over all, one and one-half inches. The shank of the anchor to intersect the hemisphere between the Arctic Circle and the Tropic of Cancer on the left, and the Equator and Tropic of Carpicorn on the right. The whole to be on a ground of dark-blue cloth.

3. Officer's Epaulette Ornaments.

   Device of the corps in silver, placed within the crescent.

4. Enlisted Full-Dress Hat Ornament.

   Ornament on front of hat, a brass United States shield, two and seven-eighths inches long by two and three eighths inches wide; resting upon the centre of the shield a nickle-plated corps device, cut from one piece of brass, of the same drsign and dimensions as prescribed for the officers full-dress hat.

5. Enlisted Undress Cap and Fatigue Cap Ornament.

   CAP ORNAMENT

   Of same design and dimensions as that prescribed for officers, but struck from one piece of brass, and fastened to the front of the cap by two brass-wire loops soldered on the inner side of the anchor.

6. Officers Dress-Coat Collar Ornaments.

(Extract from General Order No. 4, HQMC, dated 29 October 1877.)

Rank to be indicated on the ends of the collar under the corps device, which shall be one-half the size prescribed for the epaulette.

## APPENDIX D

## THE 1892 UNIFORM ORNAMENTS

(Extracts from Regulations for the Unifor m and Dress, Equipments, Etc., Of The United States Marine Corps, July 14, 1892.)

1. Officer's Black Helmet Ornament.

   Device. - The same pattern as adopted for enlisted men, with fretted silver Western Hemisphere, with chased parallels; continent cf l,vr th and South America, and anchor to be of gold, eagle to be of silver.

2. Officer's Undress Cap and White Cap Ornaments.

### CAP ORNAMENT

A fretted silver western hemisphere, struck from solid plates with chased parallels; continents of North and South America of gold plate. To be surmounted by a spread eagle cut from solid silver, and securely fastened by a tag of silver soldered on the inner side of the hemisphere. The hemisphere to rest upon the shank , between the stock and flukes of a foul anchor in gold metal; the inner edges of the two nearly touching the hemisphere. The dimensions to be as follows: hemisphere, three-fourths of an inch in diameter ; the eagle, three-eighths of an inch from top of head to point of claws; spread of wings, one inch; width of anchor stock and shank, one-ighth of an inch, arms of anchor , in- cluding flukes, three-fourths of an inch in length from peak of flu{{es to apex of crown, and one-eighth of an inch in width; length of anchor over all, one and one-half inches. · The shank of the anchor to intersect the hemisphere between the Arctic Circle and the Tropic of Cancer on the lef t, and the Equator and Tropic of Capricorn on the right.

3. Officer's Undress Coat Collar Ornaments.

   The Corps device and insignia of rank, both of metal and one-half the size adopted for epaulettes, the rank device to be backed with scarlet cloth, to be worn on each side of the collar• . •

4. Officer's Epaulette Ornaments.

   Corps device in silver , placed within the crescent.

5. Enlisted Black Helmet Ornament.

   Device. - An enlarged Corps device, consisting of the eagle, globe, and anchor; to be made of brass, according to the standard device in the Quartermaster's office.

6. Enlisted Undress Cap Ornament.

### ORNAMENT

Of the same pattern and dimensions as the present standard sample.

7, Enlisted Shoulder Knot Ornaments,

To have , •.• a nickle plated Corps device on the center of the (shoulder knot) pad,

8. Enlisted Members of the Band Undress Coat Collar Ornaments.

a. Second leader. - , , . to have a small Corps device of white metal near each point of the collar,

b, Musicians, - , , , , nPar each point (of the collar) , a small Corps ,device of white metal,

## APPENDIX E

## THE 1900 UNIFORM ORNAMENTS

(Extracts from Regulations Governing the Uniform and Equipments of Officers and Enlisted Men of the United States Marine Corps, 1900)

1. Brigadier General, Commandant and Staff Officer's Chapeau Ornament.

   On the right side to have an ornament consisting of a gold lace strip two and one-fourth inches wide with embroidered edge, on which will be placed a gilt and burnished Marhe . Corps cap ornament and a large-sized Marine Corps button.

2. Officer's Helmet Ornament.

   Corps device. - The same pattern as adopted for enlisted men, wit:1 fretted silver Western Hemisphere, with chased parallels; continent of North and South America, and anchor to be of gold, eagle to be of silver.

3. Officer's Undress Cap and Campaign Hat Ornament.

   CAP ORNAMENT

   A fretted silver western hemisphere, struck from solid plates with chased parallels; continents of North and South America of gold plate. To be surmounted by a spread eagle cutfrom solid silver, and securely f astened by a tag of silver soldered on the inner side of the hemisphere. The hemisphere to rest upon the shank, between the stock and flukes, of a foul anchor in gold metal_; the inner edges of the flukes nearly touching the hemisphere. The dimensions to be as follows: hemi- sphere, three-fourths of an inch in diameter; the eagle, three-eighths of an inch from top of head to point of claws; spread of wings, one inch; width of anchor stock and shank, one-eighth of an inch; arms of anchor, including flukes, three-fourths of an inch in length from peak of flukes to apex of crown, and one-eighth of an inch in width; length of anchor over all, one and one-half inches. The shank of the anchor to intersect the hemisphere between the Arctic Circle and the Tropic of Cancer on the lef t, and the Equator and Tropic of Capricorn on the right.

4. Officer's Undress, Summer Undress and Campaign Coats Collar Ornaments.

   The Corps device and insignia of rank one-half the size adopted for epaulettes• • • • to be worn on each side of the collar• • • •

5. Officer's Epaulette Ornaments.

   Corps device in silver placed within the crescent• • •

6. Staff Officer's Shoulder Knot Ornaments.

And a Corps device of gold and silver, of size midway between the collar and cap ornaments, to be worn on end of knot, flukes of anchor pointing to the front• • •

7. Enlisted Helmet Ornament.

Device. - An enlarged Corps device, consisting of the eagle, globe, and anchor, to be made of brass, according to the standard sample in the Quartermaster's Department. All ornaments shall be fire gilt.

8. Enlisted Undress Cap and Campaign Hat Ornament.

ORNAMENT

A Corps device of gilders' metal No. 14, of the same pattern and dimensions as the present standard sample in the Quartermaster's Department.

9. Enlisted Shoulder Knot Ornaments.

• • • same as standard sample iri the Quartermaster's Department.

# APPENDIX F

## THE 1904 AND 1908 UNIFORM ORNAMENTS

(Extracts from Regulations Governing the Uniform and Equipments of Officers and Enlisted Men of the United States Marine Corps, March 16., 1904.

and

Regulations Governing the Uniform and Equipments of Officers and Enlisted Men of the United States Marine Corps, 1908.

The sections pertaining to the ornaments are identical, subject to minor modifications, in both sets of regulations.)

1. *Brigadier General, Commandant and Staff Officer's Chapeau Ornament.

   On the right side to have an ornament consisting of a gold-lace strip 2 1/.4 inches wide, with embroidered edge, on which will be placed a gold and burnished Marine Corps cap ornament and a large-size Marine Corps button.

2. Officer's Full-dress, Undress and White Cap Ornament.

   CORPS DEVICE FOR CAPS

   A fretted silver Western Hemisphere, struck from solid plates with chased parallels; continents of North and South America in gold. To be surmounted by a spread eagle cut from solid silver, and hard-soldered to the hemisphere; eagle to face away from anchor ring. The hemisphere to rest upon the shank, between the stock and flukes of a foul anchor of smooth gold; the inner edges of the flues to be distant 1/16 inch from the hemisphere. The dimensions to be as follows: Hemisphere, 7/8 inch in diameter, the eagle, 1/2 inch from top of head to point of claws; spread of wings, 1 3/16 inches; width of anchor stock and shank, 1/8 inch; arms of anchor, from peak of flukes to apex of crown, 13/16 inch; length of anchor over all, 13/4 inches. The shanks of the anchor to intersect the hemisphere between the Arctic Circle and the Tropic of Cancer on the left and the Equator and Tropic of Capricorn on the right. **To conform to standard sample in the office of the Quartermaster, U. S. Marine Corps

3. Officer's Khaki Cap and Field Hat Ornament.

   CORPS DEVICE FOR **KHAKI** CAPS

   A device similar in all respects as to shape, size, and design to that stated above, but made in dull-finished bronze metal.

*"Brigadier General" in the 1904 Regulations; "Major General" in the 1908 Regulations•.

**The terminology "To conform to standard sample in the office of the Quartermaster, U. S. Marine" was eliminated from the 1908 Regulations.

4. Officer's Blue and White Undress Coats and White Mess Jacket Collar Ornaments,

CORPS DEVIC'::: OR COLLARS OF UNDR ESS COATS

Afretted silver Western Hemisphere, struck from solid plates with chased parallels; continents of North and South America in gold, To be surmounted by a spread eagle cut from solid silver , and hard-soldered to the hemisphere, eagle to face away from anchor ring. the The hemisphere to rest upon shank, between the stock and flukes of an anchor of smooth gold; the inner edges of the flukes distant 1/16 inch from the hemisphere. The dimensions to be as follows: Hemisphere, 11/16 inch in diameter; the eagle, 7/16 inch from the top of the head to point of claws; spread of wings, 1 1/16 inches; width of anchor stock, 3/32 inch; shank, full 1/16 inch; arms of anchor , from peak of flukes to apex of crown, 10/16 inch; length of anchor over all, 1 3/8 inches• . The shank of the anchor to intersect the hemisphere between the Arctic Circle and the Tropic of Cancer on the left, and the Equator and Tropic of Capricorn on the right, **To conform to standard sample in the office of the Quartermaster , U. S. Marine Corps.

5, Officer's Field Coat Collar Ornaments

COR PS DEVICE FOR COLLARS OF FIE LD COATS

A device similar in all respects as to shape, size, and design to that for collars of undress coats, but made of dull-finished bronze metal.

6, Officer's Epaulette Ornaments

COR PS DEVICE FOR E PAULE TTES

A fretted silver Wester n H emisphere, struck from solid plates with chased par allels; continents of North and South America in silver. To be surmounted by a spread eagle cut from solid silver and hard- soldered to the hemispher e; eagle to face away from the anchor ring. The hemisphere to rest upon the shank, between the stock and flukes, of an anchor of smooth silver metal ; the inner edges of the flukes to be distant 1/16 inch from the hemisphere. The dimensions to be as follows: Hemisphere , 7/8 inch in diameter ; the eagle, 1/2 inch from top of head to point of claws; spread of wings, 1 3/16 inches; width of anchor stock and shank , 1/8 inch; arms of anchor , from peak of flukes to apex of crown 13/16 inch; length of anchor over all, 13/4 inches. The shank of the anchor to intersect the hemisphere between the Arctic Circle and the Tropic of Cancer on the lef t, and the Equator and Tropic of Capricorn on the right. **To conform to standard sample in the office of the Quar termaster , U. S. Marine Corps.

**The terminology "To conform to standard sample in the office of the Quartermaster , U. S. Marine Corps" was eliminated from the 1908 Regulations.

7, Staff Officer's Shoulder Knot Ornaments,

CORPS DEVICE FOR SHOULDER KNOTS

To be all silver and of same size as collar ornament for undress coat. **To conform to standard sample in the office of the Quarter- master, U, S, Marine Corps.

8, Enlisted Full-dress Cap, Dress (Blue and White) Caps and Field Hat Ornament (1904 Uniform Regulations).

ORNAMEN1

A Corps device of gilder's metal No, 14, of the same pattern and dimensions as the present standard sample in the Quartermaster's Department, will be worn on the front part of full-dress cap, dress cap (white and blue), and field hat,

9, Enlisted Full-dress Cap, Dress (Blue and White) Cap and Field Hat Ornaments (1908 Uniform Regulations).

A Corps device of gilder's metal No, 14, of the same pattern and dimensions as the present standard sample in the Quartermaster's Office, **will** be worn on the front of full-dress cap and undress cap (white and blue), A Corps device of the same pattern and dimensions in· dl,lll-finished bronze **will** be worn on the field hat.

**The terminology "To conform to standard sample in the office of the Quarter-master, U. S. Marine Corps" was eliminated from the 1908 Regulations.

APPE NDIX G

THE 1912 AND 1917 UNIFOR M ORNAME NTS

(Extracts from Uniform Regulations United States Marine Corps, 1912

and

Cniform Regulations United States Marine Corps 1912 , Revi sed 1917,

The sections pertaining to the ornaments are identical in both sets of regulations,)

1. Major Gener al Comm andant and Staff Officer's Chapeau Ornament,

   On the upper end of the strap there shall he a corps device of gold of the same size as the device presc ribed for caps, and at the bottom of the strap a 40-ligne Marine Corps gilt hutton,

2, Officer's Full-dress, Undress and White Cap Ornament, Leader of the Band Special Full-dress, Full-dress and Dress Cap Ornament,

   COR PS DE VICE FOR FU LL-DR ESS AND UNDR ESS CAPS

   The corps device for the full-dress and undress caps shall be a f retted silver Western Hemisphere struck from solid plate with chased par allels, and the continents of North and South Amer ica in gold, resting upon the shank between the stock and flukes of a foul anchor of smooth gold; the inner edges of the flukes distant 1/16 inch from the hemisphe re, and surmounted by a spread eagle cut from solid silver and hard soldered to the hemi sphere, with the eagle facing away from the anchor ring. The dimensions shall he as follows: Hemisphere , 7/8 of an inch in di ameter , the eagle 1/2 of an inch from the top 9f the head to the point of the claws; the spread of the wings, 1 3/16 inches ; the width of the anchor stock and shank, 1/8 of an inch; the arms of the anchor , from the peak of the flukes to the apex of the crown , 13/16 of an inch ; the length of the anchor over all, 1 3/4 inches. The shank of the anchor shall intersect the hemisphere at an angle of 30 degrees with the plane of the Equator. On the hack of the device there shall he a screw post fitted with a milled nut for securing the device to the cap through an eyelet at the proper place.

3. Officer's Summer and Wi nter Field Caps and Field Hat Ornament,

   COR PS DEVICE FOR FIE LD CAPS AN D HATS

   The corps device for field caps and hats shall be similar in all respects as to shape, size, and design to that presc ribed for f ull-dress and undress cap , and shall be made of dull-f inish bronze metal,

4. Officer's undress and White Undress Coats and White Mess J acket Collar Ornaments.

Leader of the Band Special Full-dress Coat Collar Ornaments.

CORPS DEVICE FOR COLLARS OF UNDRESS COATS

The corps device for the coliars of undress coats shall consist of a fretted silver Western Hemisphere struck from a solid plate, with chased par allels, and the continents of North and South America in gold, resting upon the shank between the stock and flukes of an anchor of smooth gold; the inner edges of the flukes distant 1/16 inch from the hemisphere, and surmounted by a spread eagle cut from solid silver and hard soldered to the hemisphere, the the eagle facing away from the anchor ring. The dimensions shall be as follows: Hemisphere, 11/16 of an inch in diameter ; the eagle, 7/16 of an inch from the top of the head to the point of the claws; the spread of the wings,l 1/16 inches; the width of the anchor stock , 3/32 of an inch; the shank, fu.l 1/16 of an inch: the arms of the anchor , from the peak of the flukes to the apex of the crown, 10/16 of an inch; the length of the anchor over all, 1 3/8 inches. The shank of the anchor shall intersect the hemisphere at an angle of 30 degrees with the plane of the Equator, These devices shall be made in pairs, rights and lef ts, and shall be worn with the point of the crown of the anchor to the front, and attached to the coat by clasp pins,

5. Officer's Summer and Winter Field Coats Collar

Ornaments. CORPS DEVICE FOR·coLLARS OF FIELD

COATS

The corps device for the collars of field coats shall be similar in all respects as to shape, size, and design to that prescribed for the collars of undress coats, and shall be made of dull-finish bronze metal,

6. Officer's Epaulet Ornaments.

CORPS DEVICE FOR EPAULETS

The corps device for epaulets shall consist of a fretted silver Western H emisphere, struck from a solid plate, with chased parallels, and the Continents of North and South America in silver, resting upon the shank between the stock and flukes of an anchor of smooth silver, and surmounted by a spread eagle cut from solid silver and hard soldered to the hemisphere, the eagle facing away from the anchor ring. The inner edges of the flukes shall be distant 1/16 of an inch from the hemisphere, The dimensions shall be as follows: Hemisphere, 7/8 inch in diameter; the eagle, 1/2 inch from the top of the head to the point of the claws; the spread of the wings, 1 3/16 inches; the width of the anchor stock and shank, 1/8 inch; the arms of the anchor from the peak of the flukes to the apex of the crown, 13/16 inch; the length of the anchor over all , 1 3/4 inches, The shank of the anchor shall intersect the hemisphere at an angle of 30 degrees with the plane of the Equator. These devices shall be made in pairs, rights and lef ts, and shall be worn with the point of the crown of the anchor to the front.

7, Enli sted Cap and Hat Ornaments,

CORPS DEVICES FOR CAPS A:\D HAT

A corps device, struck from one piece of gilder's metal Ko, 14, of the same pattern and dimensions as the standard sample in the Quarter- master's Department of the Il arine Corps, shall be worn on the front of the dress cap, When the white cap cover is worn, this device shall be worn outside of the cover, A corps device of the same pattern and dimensions in dull-finish bronze shall be worn on the field cap and field hat. The dimensions of these devices shall be the same as prescribed for officers,

8. Enlisted Collar Ornaments

(Extracts from Changes o. 20 of 25 February 1920 to Uniform Regulations Cnited States Marine Corps, 1912, Revised 1917,)

COLLAR OR A:\IEKTS

All enlisted men, - The collar ornaments, similar in design to and approximately two-thirds of the size of those prescribed for officers, shall be made according to the standard sample in the Quartermaster's Department of the i\I arine Corps, They shall be worn vertically in the 9enter of each side of the collar, crown of anchor to the front, center of ornament 1 1/2 inches from neck opening. Gilt ornaments shall be worn on dress coat, and dull-finish bronze ornaments on service coats, summer and winter,

Second leader of the band and musicians, - The collar ornaments shall be the same as prescribed for other enlisted men of the :Marine Corps and worn in the same position on the coats, Gilt ornaments shall be worn on special full-dress, full-dress, and dress coats, and dull-finish bronze ornaments on service coats, summer and winter.

Drum major, - The collar ornaments shall he the same as prescribed for other enlisted men of the :\larine Corps and worn in the same position on the coats. Gilt ornaments shall be worn on full-dress and dress coats, and dull-finish bronze ornaments on service coats, st;mmer and winter.

9, Revisions in the 1917 Cniform Regulations as a Result of the Addition of the Garrison Cap as a Part of the Cniform.

(Extracts from Changes o. 22 of 9 October 1920 to Cniform Regulations Cnited States Marine Corps 1912, Revised 1917.)

a. Officer's Summer and Winter Field Caps, Garrison Cap and Field Hat Ornament.

## CORPS DEVICE FOR FIELD CAPS, GARRISON CAP AND HAT

The corps device for field caps, the garrison caps, and field hat shall be similar in all respects as to shape, size and design to that prescribed for dress and undress caps, and shall be made of dull-finish bronze metal.

b. Enlisted Summer and Winter Field Caps, Garrison Cap and Field Hat Ornament.

## CORPS DEVICES FOR CAPS AND HATS

A corps device, struck from one piece of gilder's metal No. 14, of the same pattern and dimensions as the standard sample in the Quartermaster's Department of the Marine Corps, shall be worn on the front of the dress cap. When the white cap cover is worn, this device shall be worn outside of the cover. A corps device of the same pattern and dimensions in dull-finish bronze shall be worn on the field cap, garrison cap, and field hat. The dimensions of these devices shall be the same as prescribed for officers.

## APPENDIX H

## THE 1922 UNIFORM ORNAMENTS

(Extracts from the Uniform Regulations United States Marine Corps 1922.)

1. Officer's Cap and Hat Ornaments.

### CAP AND HAT ORNAMENTS

All officer's. - The gold and silver cap ornament shall be worn on the dress and undress caps, and the bronze cap and hat ornament on the winter and summer service caps, the garrison cap, and the field hat, being worn on the garrison cap on the left front side, with the center of the ornament 2 inches from the front.

2. Officer's Dress and Undress Cap Ornament.

Leader of the Band Special Full-dress and Full-dress Cap Ornament.

### ORNAMENT, CAP, GOLD AND SILVER

To consist of Western Hemisphere, intersected by a foul anchor, and surmounted by an eagle. To be according to standard sample in the Quartermaster's Department.

Hemisphere. - To be perfect hollow half globe, of fretted silver struck from a solid plate, seven-eighths inch in diameter, with accurately proportioned and clearly outlined continents of North and South America (in gold) in raised surfaces, and be chased with nine parallels of latitude, terminating at continents.

Eagle, with outspread wings, cut from solid silver, to stand on top of hemisphere facing away from anchor ring, and be hard soldered to globe. To be one-half inch from crest of head to point of claws; from tip to tip of wings, 1 3/16 inches. Eagle to be accurately proportioned, with claws and head, and feathers of neck, body, wings, and legs distinctly delineated.

Foul anchor. - To be of smooth gold, the shank between stock and flukes to intersect hemisphere from northeast to southwest at an angle of 30 degrees with plane of the Equator. Width of stock and shank, one-eighth inch; length of stock, five-eighths inch; distance peak of flukes to apex of crown, thirteen-sixteenths inch; inner edge of hemi- sphere to top of anchor shackle or ring, one-half inch, and to apex of crown, three-eighths inch; over-all length of anchor, 1 3/4 inches. Cable to be attached to ring and twine around shank and arms. Each detail of anchor to be clearly delineated.

Spike and screw post. - A thin spike, three-sixteenths inch long, tapering to point, to project from center of shoulders of eagle. A threaded screw post, with washer and milled nut, for securing ornament through eyelet to dress caps, undress caps, and full-dress and special full-dress cap, leader , Marine Band, to project from center of hollow globe and be securely soldered thereto,

3. Officer's Summer and Winter Service Caps, Garrison Cap and Field Hat Ornament.

ORNAME NT, CAP AND HAT, BRONZ E

To consist of Western Hemisphere , intersected by a foul anchor and surmounted by an eagle, To be made of dull-finish bronze metal and be according to standard sample in the Quartermaster's Department,

Hem isphere. - To be perf ect hollow half globe, seven-eighths inch in diameter, with accurately proportioned and clearly outlined continents of North and South America in raised surfaces, and be chased with nine parallels of latitude, terminating at continents.

E agle , with outspread wings, to stand on top of hemisphere facing away from anchor ring, and be hard soldered to globe, To be one-half inch from crest of head to point of claws; from tip to tip of 'wings, 1 3/16 inches. E'agle to be accurately propor tioned, with c l a w and head , and feathers of neck , ody wings, and legs distinctly delineated.

Foul Anchor, - The shank between stock and flukes to intersect hemisphere from northeast to southwest at an angle of 30 degrees with plane of the Equator. Width of stock and shank , one-eight inch; length of stock, five- eighths inch; distance peak of flukes to apex of crown , thirteen-sixteenths inch; inner edges of flukes, one sixteenth inch from hemisphere; distance edge of hemisphere to top of anchor shackle or ring, one-half inch , and to apex of crown , three-eighths inch; over-all length of anchor , 1 3/4 inches. C able to be attached to ring and twine around shank and arms, E ach detail of anchor to be clearly delineated.

Spike and screw post. - A thin spike three sixteenths inch long, tapering to point, to project from center of shoulders of eagle. A threaded screw post, with washer and milled nut, for securing ornament through eyelet to service and garrison caps and field hat to project from center of hollow globe and be securely soldered thereto,

4. Officer's Collar and Shoulder-knot Ornaments.

COLLAR OR NAME NTS AND COLLAR INSIGNIA

Collar ornaments ( all officers) . - The collar ornaments shall be worn on each coat, except the evening dress jacket, by all officers, being placed vertically in the center of each side of the collar, with eagle facing toward the front, and center of ornament 1 1/2 inches from the neck opening.

The gold and silver collar ornaments are presc ribed for the collars of the dress and undress coats, and the bronze collar ornaments are presc ribed for the collars of the winter and summer service coats.

Commissioned officers. - The shoulder knots worn with the evening dress jacket shall have the following placed thereon: In the case of the line officers, the gold and silver ornaments prescribed. • • in the case of permanent staff officers, and those detailed to a staff department for a four-year period, the respective gold and silver staff insignia in addition to the ornament and insignia of rank• • •

The ornament and insignia shall be worn on center line of shoulder knot, equally spaced between lower end of knot and the button at collar end. The ornament shall be placed at shoulder end of knot, with head of eagle toward collar end, eagle facing the front• • •

5. Officer's Dress and Undress (White) Coats and Shoulder-knot Ornaments.

ORNAMENTS, COLLAR AND SHOULDER-KNOT, GOLD AND SILVER

To consist of Western Hemisphere, intersected by an anchor , and surmounted by an eagle. To be made in pairs, right and left, and be according to standard sample in the Quartermaster's Department.

Hemisphere. - To be perfect hollow half globe of fretted silver struck from a solid plate, eleven-sixteenths inch in diameter , with accurately proportioned and clearly outlined continents of North and South America (in gold) in raised surfaces, and be chased with nine parallels of latitude, terminating at continents.

Eagle, with outspread wings, cut from solid silver , to stand on top of hemisphere facing away from anchor ring, and be hard soldered to globe. From crest of head to point of claws, seven-sixteenths inch; from tip to top of wings, 1 1/16 inches. Eagle to be accurately proportioned, with claws and head, and feathers of neck , body , wings, and legs distinctly delineated.

Anchor. - To be of smooth gold, the shank between stock and flukes to intersect hemisphere at an angle of 30 degrees with plane of the Equator. The anchor on left ornament to intersect hemisphere from northeast to southwest and that on right ornament from northwest to southeast. Width of stock three Thirty seconds inch, and of shank full one-sixteenth inch; length of stock, one-half inch; distance peak of flukes to apex of crown, five-eighths inch;inner edges of flukes one-sixteenth inch from hemisphere; distance edge of hemisphere to top of anchor shackle or ring, three-eighths inch , and to apex of crown, three-eighths inch; over-all length of anchor 1 3/8 inches.

Fastening. - To have vertical clasp pin or screw fastening for attaching ornament to dress and undress coats and shoulder knots.

6. Officer's Summer Service and Winter Service Coats Collar Ornaments.

## ORNAMENTS, COLLAR, BRONZE

To consist of Western Hemisphere, intersected by an anchor, and surmounted by an eagle. To be made of dull-finish bronze metal, in pairs, right and left, and be according to standard sample in the Quartermaster's Department.

Hemisphere. - To be perfect hollow half globe struck from a solid plate, eleven-sixteenths inch in diameter, with accurately proportioned and clearly outlined continents of North and South America in raised surfaces, and be chased with nine parallels of latitude, terminating at continents.

Eagle, with outspread wings, to stand on top of hemisphere facing away from anchor ring, and be hard soldered to globe. From crest of head to point of claws, seven-sixteenths inch; from tip to tip of wings, 1 1/16 inches. Eagle to be accurately proportioned, with claws and head, and feathers of neck, body, wings, and legs distinctly delineated.

Anchor. - The shank between stock and flukes to intersect hemisphere at an angle of 30 degrees with plane of the Equator. The anchor on left ornament to intersect hemisphere from northeast to southwest and that on right ornament from northwest to southeast. Width of stock three thirty-seconds inch and shank full one-sixteenth inch; length of stock, one-half inch; distance peak of flukes to apex of crown, five-eighths inch; inner edges of flukes, one-sixteenth inch from hemisphere; distance edge of hemisphere to top of anchor shackle or ring, three-eighths inch, and to apex of crown three-eighths inch; over-all length of anchor, 1 3/8 inches.

Fastening. - To have vertical clasp pin or screw fastening for attaching ornament to service coats.

7. Enlisted Cap, Hat and Collar Ornaments.

### CAP AND HAT ORNAMENTS

Enlisted men. - The gilt cap ornament shall be worn by all enlisted men on the dress and white caps, and the bronze cap and hat ornament on the winter and summer service caps, the garrison cap, and the field hat. On the garrison cap, the bronze cap and hat ornament shall be worn on the left front side, in the place provided therefor.

### COLLAR ORNAMENTS

Enlisted men. - The collar ornaments shall be worn by all enlisted men on the collars of all coats, being placed vertically in the center of each side of the collar, with eagle facing the front and center of ornament 1 1/2 inches from the neck opening.

Enlisted men. - Gilt collar ornaments are prescribed for the collar of the dress coat, and bronze collar ornaments for the winter and summer service coats.

8. Marine Band Cap, Hat, Collar and Shoulder-knot Ornaments.

INSIGNIA

Leader. - On the shoulder knot the insignia embroidered on scarlet cloth and the gold and silver shoulder-knot ornament shall be placed on the center line, equally spaced between lower end of knot and the button at collar end, the ornament being placed at shoulder end of knot, with head or' eagle toward collar end, eagle facing the front.

CAP AND HAT ORNAMENT

Leader. - The gold and silver ornament shall be worn on the special full-dress, full-dress, dress, and undress cap, and the bronze cap and hat ornament, on the winter and summer service caps and the field hat.

Second leader , drum major , and musicians. - The gilt cap ornament shall be worn on the special full-dress (except drum major) , full-dress, dress, and white caps, and the bronze cap and hat ornament on the winter and summer service caps and the field hat.

COLLAR ORNAMENTS

The collar ornaments shall be placed vertically in the center of each side of the collar of the coats indicated, with eagle facing the front, center of ornaments being 1 1/2 inches from the neck opening.

Leader. - Gold and silver ornaments, are prescribed for the collars of the dress and undress coats, and bronze collar ornaments for the winter and summer service coats.

Second leader , drum major , and musicians. - Gilt collar ornaments are prescribed for the collars of the special full-dress (except drum major) , full-dress, and dress coats, and bronze collar ornaments for the winter and summer service coats.

9. Officer's Col.lar Ornaments for the Roll-Collar t ummer Service and Winter Service Coats.

(Extracts from Marine Corps Circular Letter No. 46 of 6 December 1926.)

DEVICES FOR ROLL-COLLAR COATS, OFFICERS

A smaller bronze collar ornament has been adopted for roll-collar coats for officers, to be obligatory on 1 July , 1927, for winter service, and 1 January 1928, for summer service, but optional in the meantime. These ornaments will be three-fourths the size of the bronze collar ornaments heretofore specified, and will be made in pairs, right and lef t, in one piece, and follow the design of the basic Marine Corps emblem, omitting the m o t , ribbon and rope.

On roll-collar coats for officers the position of the ornaments on the collar is changed. and insignia worn by Warrant Officer's will be worn on the lapel instead of the collar. paragraph 3 of reference (b) being modified accordingly. The or.narnents will be centered on the collar between the crease roll and the outer edge. center of hemisphere 1 inch above the horizon tal edge of notch, and the insignia will be placed on the lapel with the center mid-way on a line bisecting the angle of the peak of lapel.

## APPENDIX I

## THE 1929 UNIFORM ORNAMENTS

(Extract.s from Uniform Regulations United States Marine Corps 1929.)

1. Officer's Cap and Hat Ornaments.

### CAP AND HAT ORNAMENTS

All officers. - The dress ornament shall be worn on the full-dress, dress, and undress caps, and the service cap and hat ornament on the winter and summer service caps, the garrison caps, and the field hat, being worn on the garrison caps on the left front side, with the center of the ornament 2 inches from the front.

2. Officer's Dress and White Undress Cap Ornament.

Leader, Marine Band and Specified Officer's Special Full-dress and Full-dress Cap Ornament.

Ornament cap, dress. - To consist of Western Hemisphere, intersected by a foul anchor, and surmounted by an eagle. Hemisphere to be perfect hollow half globe of fretted silver, with continents of North and South America in gold, eagle to be cut from solid silver, and foul anchor to be of smooth gold; design and measurements to be as specified for hemisphere, eagle, and foul anchor, respectively, of service cap and hat ornament. Screw post, with washer and nut, for securing ornament through eyelet to dress cap, white undress cap, and full-dress and special full-dress cap, for officers specified and leader, Marine Band, also to be as indicated on service ornament. According to standard sample in the Quartermaster's Department.

3. Officer's Summer and Winter Service Caps, Garrison Caps and Field Hat Ornament.

Ornament, cap and hat, service. - To be made of dull-finish bronze metal throughout, and to consist of Western Hemisphere, intersected by a foul anchor, and surmounted by an eagle. To be according to standard sample in the Quartermaster's Department.

Hemisphere. - To be perfect hollow half globe, 7/8 inch in diameter, with accurately proportioned and clearly outlined continents of North and South America in raised surfaces, and be chased with nine parallels of latitude, terminating at continents.

Eagle, with outspread wings, to stand on top of hemisphere facing away from anchor ring, and be hard soldered to globe. To be 1/2 inch from crest of head to point of claws; from tip to tip of wings, 13/16 inches. Eagle to be accurately proportioned, with claws and head, and feathers of neck, body, wings, and legs distinctly delineated. A thin spike, 3/16 inch long, tapering to point, to project from center of shoulders of eagle.

Foul anchor. - The shank between stock and flukes to intersec t hemisphere from northeast to southwest at an angle of 30 degrees with plane of the E quator. Wid th of stock and shank , 1/8 inch; length of stock, 5/8 inch; distance peak of flukes to apex of crown , 13/16 inch; inner edges of flukes 1/16 inch from hemisphere; distance edge of hemisphere to top of anchor shackle or ring, 1/2 inch , and to apex of crown , 3/8 inch; over-all length of anchor , 1 3/4 inches. C able to be attached to ring and twine around shank and arms. E ach detail of anchor to be clearly delineated,

Screw post. - A threaded screw post, with washer and milled nut for securing ornament through eyelet to service and garrison caps and field hat, to projec t from center of hollow globe and be securely soldered thereto.

4. Officer's Collar and Shoulder Knot Ornaments.

COLLAR ORNAM ENTS

All off icers. - The collar ornaments shall be worn on the dress, white undress, and service coats, and the whi te mess jacket, On the standing ..::ollar

dress and und ress coats and whi te mess jacket, the ornaments shall be ver placed tically in the center of each side of the collar , with the eagle toward facing the front, center of hemisphere of ornament being 1 1/2 inches from the neck opening. On the roll collar service coats, the ornaments shall be worn on the collar , centered between the crease roll and outer edge, wi th center of hemisphere 1 inch above the horizontal edge of lapel notch, the vertical axis of ornament being parallel with crease roll and the eagle facing inside, Collar ornaments shall not be worn on the cotton and flannel shirts.

The dress collar ornaments are presc ribed for the collars of the dress and und ress coats and white mess jacket, and the service collar ornaments are presc ribed for the collars of the winter and summer service coats,

ORNAM ENTS AND INSIGNIA FOR SHOULDE R K NOTS

Commissioned off icers. - The shoulder knots worn with the full-dress coat and the evening dress jacket shall have the following placed thereon: In the case of line officers, the dress ornaments prescribed . . .and the insignia of rank embroidered on scarlet cloth; in the case of permanent staff officers, and those detailed to a staff department for a four-year period, the respec- tive dress staff insignia• .,in addition to the ornament and insignia of rank• . .

The ornament and insignia shall be worn on center line of shoulder knot, equally spaced between lower end of knot and the button at collar end, The ornament shall be placed at shoulder end of knot, with head of eagle toward collar end , eagle facing the front• • •

5. Officer's Dress and Undress (White) Coats, White Mess Jacket and Shoulder Knot Ornaments,·

Ornaments, collar and shoulder knot, dress. - To consist of Western Hemisphere, intersected by an anchor, and surmounted by an eagle. To be made in pairs, right and lef t, facing of eagle and direction of anchor being changed. To be· according to standard sample in the Quartermaster's Department.

Hemisphere. - To be perf ect hollow half globe of fretted silver struck from a solid plate 11/16 inch in diameter , with accur ately propor tioned and of clearly outlined continents North and South America (in gold) in raised with surfaces, and be chased nine parallels of latitude, terminating at continents.

E agle, wlth outspread wings, cut from solid silver , to stand on top of hemisphere facing away from anchor ring, and be hard soldered to globe. From crest of head to point of claws, 7/16 inch; from tip to tip of wings, 11/16 inches. E agle to be accurately proportioned, with claws and head, and f eathers of neck , body , wings, and legs distinctly delineated.

Anchor. - To be of smooth gold, the shank between stock and flukes to intersect hemisphere at an angle of 30 degrees with plane of the Equator. The anchor on lef t ornament to intersect hemisphere from northeast to south- west and that on right ornament from northwest to southeast. Width of stock , 3/32 inch, shank full, 1/16 inch; length ot stock, 1/2 inch; distance peak of flukes to apex of crown, 5/8 inch; inner edges of flukes, 1/16 inch from hemisphere; distance edge of hemisphere to top of anchor shackle or ring, 3/8 inch , and to apex of crown, 3/8 inch; over-all length of anchor , 1 3/8 inches.

Fastener. - To have vertical clasp pin, or screw or clutch fastener for attaching ornament to dress and undress coats, white mess jacket, and shoulder knots.

6. Officer's Summer Service and Winter Service Coats Collar Ornaments.

Ornaments, collar , service. - To be made in one piece of dull-finish bronze metal throughout. To be in same design as standard Marine Corps emblem (omitting motto ribbon and anchor rope) , and consist of Western Hemisphere, intersected by an anchor , and surmounted by an eagle. To be in pairs, right and lef t, facing of eagle and direction of anchor being changed. To be according to the standard sample in the Quartermaster's Department.

Hemisphere. - To be perf ect hollow half globe, 23/40 inch in diameter , with accurately proportioned and clearly outlined continents of North and South America in raised surfaces , and be chased with 9 parallels of longitude and 13 parallels of latitude, terminating at continents.

E agle, with outspread wings, to stand on top of hemisphere f acing away from anchor ring. From crest of head to point of claws, 7/20 inch; from tip to tip of wings, 7/8 inch. E agle to be accurately proportioned, with claws and head, and feathers of neck, body, wings, and legs distinctly delineated.

Anchor. - The shank between stock and flukes to intersect hemisphere at an angle of 30 degrees with plane of the Equator, The anchor on lef t ornament to intersect hemisphere from northeast to southwest and that on right ornament from northwest to southeast. Width of stock, 1/10 inch and of shank, full 1/20 inch; length of stock, 17/40 inch; distance peak of flukes to apex of crown, 9/20 inch; distance edge of hemisphere to top of anchor shackle or ring, 9/40 inch, to inner edges of flukes, 1/40 inch, and to apex of corwn, 3/10 inch; over-all length of anchor 13/40 inches.

Fastener. - To have vertical clasp pin, or screw or clutch fastener, as may be most suitable for attaching ornaments to service coats.

7, Enlisted Cap, Hat and Collar Ornaments.

CAP AND HAT ORNAMENTS

Enlisted men. - The gilt cap ornament shall be worn by all enlisted men on the dress and white caps, and the bronze cap and hat ornament on the win ter and summer service caps, the garrison c·aps, and the field hat. On the garrison caps, the bronze cap and hat ornament shall be worn on the lef t front side, in the place provided therefor.

COLLAR ORNAMENTS

Enlisted men. - The collar ornaments shall be worn by all enlisted men on the collars of all coats. On t h standing collar dress coat, the ornaments shall be placed vertfcally in the center of each side of the collar, with eagle facing towarq the front, center of hemisphere of ornaments being 1 1/2 inches from t e neck opening. On roll collar service coats, the orna- men ts shall be placed with the center of hemisphere 7/8 inch above the lapel notch and 1 ich from outside edge of collar, the vertical axis of orna- men t being at right angles to the horizontal edge of notch, with eagle facing inside.

Gilt collar ornaments are prescribed for the collar of the dress coat and bron ze collar ornaments for tqe winter and summer service coats.

8. Marine Band Cap, Hat, Collar and Shoulder Knot Ornaments.

INSIGNIA

Leader. - On the shoulder knot, the insignia embroidered on scarlet cloth and the dress shoulder knot ornament shall be placed on the center line, equally spaced between lower end of knot and the button at collar end, the ornament being placed at shoulder end of knot.

CAP AND HAT ORNAME NTS

Leader. - Tpe dress cap ornament shall be worn on the special full-dress, full-dress, dress, and undress caps, and the service cap and hat ornament on the winter and summer service caps, and the field hat.

Second leader, drum major, and musicians. - The gilt cap ornament shall be worn on the special full-dress (except drum major), full-dress, dress, and white caps, and the bronze cap and hat ornament on the winter and summer service caps and the field hat.

## COLLAR ORNAMENTS

On the standing collar coats indicated, the collar ornaments shall be placed vertically in the center of each side of the collar, with eagle facing toward the front, center of hemisphere of ornament being 1 1/2 inches from the neck opening. On roll collar service coats, the ornaments shall be worn on the collar, with eagle facing inside. The ornaments worn on service coats by the leader shall be centered between the crease roll and outer edge, with center of hemisphere 1 inch above the horizontal edge of lapel notch, the vertical axis of ornament being parallel with crease roll. The ornaments worn thereon by the second leader, drum major, and musicians shall be placed with the center of hemisphere 7/8 inch above the lapel notch and 1 inch from outside edge of collar, the vertical axis of ornament being at right angles to the horizontal edge of notch, with eagle facing inside.

Leader. - The dress collar ornaments, are prescribed for the collars of the dress and undress coats, and the service collar ornaments, for the winter and summer service coats.

Second leader, drum major, and musicians. - Gilt collar ornaments are prescribed for the collars of the full-dress (except drum major) and dress coats, ancl bronze collar ornaments for the winter and summer service coats.

9. Changes to sections on Officer's Dress and White Undress Cap Ornament; Leader Marine Band and Specified Officer's Special Full-dress and Full-dress Cap Ornament; Officer's Collar and Shoulder Knot Ornaments; Enlisted Cap, Hat and Collar Ornaments.

Addition of sections on Officer's Epaulet Ornaments.

(Extracts from Changes No. 1 of 30 June 1932 to Uniform Regulations United States Marine Corps, 1929.)

a. Officer's Dress and White Undress Cap Ornament; Leader, Marine Band and Specified Officer's Special Full-dress and Full-dress Cap Ornament. (Par. 2 above.)

Ornament, cap, dress. - To consist of Western Hemisphere, intersected by a foul anchor, and surmounted by an eagle. Hemisphere to be perfect hollow half globe of fretted silver, with continents of North and South America in gold, eagle to be cut from solid silver, and anchor to be of smooth gold, with rope of silver; design and measurements to be as specified for hemisphere, eagle, and foul anchor, respectively, of service cap and hat ornament.

Screw post, with washer and nut, for securing ornament through eyelet to dress cap, white undress cap, and full-dress and special full-dress cap, for officers specified and leader, Marine Band, also to be as indicated on ser\ . . ornament. According to standard sample in the Quartermaster's Department.

b. Officer's Collar and Shoulder Knot Ornaments (Par. 4 above.)

COLLAR ORNAMENTS

To sixth line, after "the collar", insert as shown in figure 82, followed by a comma.

ORNAMENTS AND INSIGNIA FOR SHOULDER KNOTS

Commissioned officers. - The shoulder knots worn with the evening dress jacket shall have the following placed thereon: In case of line officers, the dress collar and shoulder knot ornaments, and the insig- nia of rank embroidered on scarlet cloth; in the case of permanent staff officers, and those detailed to a staff department for a four-year period, the respective dress staff insignia in addition to the ornament and insignia of rank• • •

c. Enlisted Cap, Hat and Collar Ornaments (Par. 7 above.)

COLLAR ORNAMENTS

Enlisted men. - The collar ornaments shall be worn by all enlisted men on the collars of all coats. On the standing collar dress coat, the ornaments shall be placed vertically in the center of each side· of the collar, with eagle facing toward the front, center of hemisphere of ornaments being 1 1/2 inches from the neck opening. On roll-collar service coats, the ornaments shall be. placed as shown in figure 82B, with the center of hemisphere 7/8 inch above the lapel notch and 1 inch from outside edge of collar, the vertical axis of ornament being at right angles to the horizontal edge of notch, with eagle facing inside.

Gilt collar ornaments are prescribed for the collar of the dress coat and bronze collar ornaments for the winter and summer service coats.

d. Officer's Epaulet Ornaments.

ORNAMENTS AND INSIGNIA FOR EPAULETS

Commissioned officers prescribed. - The epaulets worn with the full-dress coat prescribed for an aide-de-camp to the President and officers detailed for duty at the White House shall have placed thereon the dress epaulet ornaments and the insignia of rank embroidered on scarlet cloth.

The ornaments and insignia shall be worn on the center line of epaulet. The ornaments shall be placed on the frog or shoulder pad, with the crest of eagle 3/16 inch above cusps of metal crescent, eagle facing the front. The insignia of rank shall be equally spaced between the ornament and the button at collar end.

Ornaments, epaulet, dress, aide-de-camp to the President and officers detailed for duty at the White House. - To be made of silver throughout and to consist of Western Hemisphere, intersected by an anchor, and surmounted by an eagle. To be made in pairs, right and left, facing of eagle and direction of anchor being changed. To be according to standard sample in the Quartermaster's Department.

Hemisphere. - To be perfect hollow half globe of fretted silver struck from a solid plate 7/8 inch in diameter, with accurately proportioned and clearly outlined continents of North and South America in raised surfaces, and be chased with nine parallels of latitude, terminating at continents.

Eagle, with outspread wings, cut from solid silver, to stand on top of hemisphere facing away from anchor ring, and be hard soldered to globe. From crest of head to point of claws, 1/2 inch; from tip to tip of wings, 1 3/16 inches. Eagle to be accurately proportioned, with claws and head, and feathers of neck, body, wings, and legs distinctly delineated.

Anchor. - To be of smooth silver, the shank between stock and flukes to intersect hemisphere at an angle of 30 degrees with plane of the Equator. The anchor on left ornament to intersect hemisphere from northeast to southwest and that on right ornament from northwest to southeast. Width of stock and shank 1/8 inch; length of stock 5/8 inch; distance peak of flukes to apex of crown, 13/16 inch; inner edges of flukes 1/16 inch from hemisphere; distance edge of hemisphere to top of anchor shackle or ring, 7/16 inch, and to apex of crown, 7/16 inch; over-all length of anchor, 1 3/4 inches.

Fastener. - To have vertical clasp pin, or screw or clutch fastener, for attaching ornament to epaulets.

## APPE NDIXJ

## THE 1937 UNIRORM ORNAM ENTS

(Extracts from Uniform Regulations United States Marine Corps 1937.)

1. Officer's Cap and Hat Ornaments.

### CA P AND HAT ORNAM ENTS

All off icers. - The dress cap ornament shall be worn on the full dress, dress, and white caps; and the service cap and hat ornament on the winter and summer service caps, the garrison caps, and the field hat. On the garrison caps, the ornament is worn on the lef t front side, with the center of the ornament 2 inches from the front.

2. Officer's Full-Dress, Dress and White Cap Ornament.

Ornament , cap , dress. - Design to be that of M arine Co:i:ps emblem, omitting motto ribbon. To consist of Western Hemisphere approximately 7/8 inch in diameter , intersected by a foul anchor , and surmounted by an eagle. Hem isphere to be perfec t hollow half globe of fretted silver , with continents of North and South Americ a in gold, eagle to be cut from solid silver , and anchor to be of smooth gold with rope of silver ; design for hemisphere , eagle , and foul anchor to be as specified for service cap and hat ornament. Screw post , with washer and nut , for securing ornament through eyelet to dress cap, to projec t from center ofhollow globe and be securely soldered thereto.

3. Officer's Summer and Wi nter Service Caps, Garrison Caps and Field Hat Ornaments.

Orn ament, cap and hat , service. - Design to be that of Marine Corps emblem , omitting motto ribbon. To be made of dull-finish bronze metal throughout , and to consist of Western H emisphere , approximately 7/8 inch in diameter , intersected by a foul anchor , and surmounted by an eagle.

H emisphere. - To be perf ect hollow half globe with accurately p ropor tioned and clearly outlined continents of Nor th and South Americ a in raised surfaces , with par allels of latitude terminating at continents.

E agle , with outspread wings, to stand on top of hemisphere facing away from anchor ring. To be hard soldered to globe. E agle to be accurately proportioned with claws and head , and feathers of neck , body, wings, and legs distinctly delineated. A thin spike, 3/16 inch long, tapering to point , to projec t from center of shoulders of eagle.

Foul anchor. - The shank between stock and throat to intersect hemisphere from northeast to southwest at an angle of about 32 degrees with plane of Equ ator. Cable to be attached to ring and twine around shank and arms. E ach detail of anchor to be clearly delineated.

Screw post. - A threaded screw post, with washer and milled ,.ut, for securing ornament through eyelet to service and garrison caps and field hat, to project from center of hollow globe and be securely soldered thereto.

4. Officer's Collar, Shoulder Knot and Epaulet Ornaments.

COLLAR ORNAMENTS

All officers. - The dress collar ornaments shall be worn on the dress and white coats, the white mess jacket, and on epaulets and shoulder knots. The service collar ornaments shall be worn on the summer and winter service coats. On the dress and white coats, the ornaments shall be placed vertically in the center of each side of the collar, center of hemi- sphere 1 1/2 inches from the neck opening with eagle facing toward the front. On the white mess jacket, the collar ornament shall be worn on the center line of the shoulder strap, 3/4 inch from the armhole seam with head of eagle facing front. On epaulets and shoulder knots, the ornaments shall be placed as indicated. • • On the service coats, the ornament shall be worn on the collar, centered between the crease roll and outer edge with center of hemisphere 1 inch above the horizontal edge of lapel notch, the vertical axis of ornament being parallel with crease roll and eagle facing inside. Collar ornaments shall not be worn on the cotton and flannel shirts.

ORNAMENTS AND INSIGNIA FOR SHOULDER KNOTS

The shoulder knots worn with the evening dress jacket shall have the following placed thereon: In the case of line officers, the dress collar ornaments, and the insignia of rank embroidered on scarlet cloth; in the case of permanent staff officers, and those detailed to a staff department for a four-year period, the respective dress staff insignia in addition to the ornament and insignia of rank• • •

The ornament and insignia shall be worn on center line of shoulder knot, equally spaced between lower end of knot and the button at the collar end. The ornament shall be placed at shoulder end of knot, with head of eagle toward collar end, eagle facing the front• . •

ORNAMENTS AND INSIGNIA FOR EPAULETS

Commissioned officers prescribed. - The epaulets worn with the full-dress coat prescribed for an aide-de-camp to the President and officers detailed for duty at the White House shall have placed thereon the dress collar ornaments, and the insignia of rank embroidered on scarlet cloth.

The ornaments and insignia shall be worn on the center line of ,epaulet. The ornament shall be placed on the frog or shoulder pad, with the crest of eagle 3/16 inch above cusps of metal crescent, eagle facing the front. The insignia ·of rank shall be equally spaced between the ornament and the button at collar end.

5. Officer's Dress and White Coats, White Mess Jacket, Shoulder Knot and Epaulet Ornaments.

Ornaments, collar, dress. - Design to be that of Marine Corps emblem, omitting motto ribbon and anchor rope. To consist of Western Hemisphere, approximately 11/16 inch in diameter, intersected by an anchor, and surmounted by an eagle. To be made in pairs, right and left, facing of eagle and direction of anchor being changed.

Hemisphere. - To be perfect hollow half globe of fretted silver. struck from a solid plate, with accurately proportioned and clearly outlined continents of North and South America (in gold) in raised surfaces, and be chased with parallels of latitude, terminating at continents.

Eagle, with outspread wings, cut from solid silver, to stand on top of hemisphere facing away from anchor ring, and be hard soldered to globe. Eagle to be accurately proportioned, with claws and head, and feathers of neck, body, wings, and legs distinctly delineated.

Anchor. - To be of smooth gold, the shank between stock and throat to intersect hemisphere at an angle of 32 degrees with plane of the Equator. The anchor on left ornament to intersect hemisphere from northeast to southwest and that on right ornament from northwest to southeast.

Fastener. - To have vertical clasp pin, or screw or clutch fastener for attaching ornament to dress and white coats, white mess jacket, shoulder knots, and epaulets.

6. Officer's Summer Service and Winter Service Coats Collar Ornaments.

Ornaments, collar, service. - To be made in one piece of dull-finish bronze metal throughout. To be in same design as Marine Corps emblem (omitting motto ribbon and anchor rope), and consist of Western Hemisphere, approximately 9/16 inch in diameter, intersected by an anchor, and surmounted by an eagle. To be in pairs, right and left, facing of eagle and direction of anchor being changed.

Hemisphere. - To be perfect hollow half globe, with accurately proportioned and clearly outlined continents of North and South America in raised surfaces. and be chased with parallels of latitude. terminating at continents.

Eagle, with outspread wings, to stand on top of hemisphere facing away from anchor ring. Eagle to be accurately proportioned, with claws and head, and feathers of neck, body. wings, and legs distinctly delineated.

Anchor. - The shank between stock and throat to intersect hemisphere at an angle of 32 degrees with plane of the Equator. · The anchor on left ornament to intersect hemisphere from northeast to southwest and that on right ornament from northwest to southeast.

Fastener. - To have vertical clasp pin, or screw or clutch fastener, for attaching ornaments to service coats.

7. Enlisted Cap, Hat and Collar Ornaments.

## CAP AND HAT ORNAMENTS

Enlisted men. - The gilt cap ornaments shall be worn by all enlisted men on the dress and white caps, and the bronze cap and hat ornament on the winter and summer service caps, the garrison caps and the field hat. On the garrison caps, the bronze cap and hat ornaments shall be worn on the left front side, in the place provided therefor.

## COLLAR ORNAMENTS

Enlisted men. - Except as hereinafter provided for the Marine Band, the collar ornaments shall be worn by all enlisted men on the collars of all coats. Gilt collar ornaments shall be worn on dress coats, placed vertically in the center of each side of the collar, with eagle facing toward the front, center of hemisphere of ornaments being 1 1/2 inches from the neck opening. Bronze collar ornaments shall be worn on service coats, placed with the center of hemisphere 7/8 inch above the lapel notch and 1 inch from outside edge of collar, the vertical axis of ornament at right angles to the horizontal edge of notch, with eagle facing inside.

8. Marine Band Cap, Hat, Collar and Shoulder Knot Ornaments.

## INSIGNIA, LEADER

On the shoulder knot, the lyre, embroidered on scarlet cloth, and the dress collar ornament, shall be placed on the center line, equally spaced between lower end of knot and the button at collar end, the ornament being placed at shoulder end of knot. The insignia and ornament shall be placed with the top of the lyre and the head of the eagle toward the collar end of knot, the eagle facing toward the front.

## CAP AND HAT ORNAMENTS

Leader. - The dress cap ornament shall be worn on the full-dress, dress, and white caps.

Second leader, drum major, and musicians. - The gilt cap ornament shall be worn on the dress and white caps.

## COLLAR ORNAMENTS

Leader. - The dress collar ornaments are prescribed for the collar of the dress coat. They shall be placed vertically in the center of each side of the collar, with eagle facing toward the front, center of hemisphere of ornament being 1 1/2 inches from the neck opening.

Second leader, drum major, and musicians. - Gilt collar ornaments are prescribed for the collars of the full-dress (except drum major) and dress coats. They shall be worn on the collars thereof as above specified.

9. Changes to sections on Officer's Cap and Hat Ornaments; Officer's Collar, Shoulder Knot and Epaulet Ornaments; Enlisted Cap, Hat and Collar Ornaments.

(Extracts from HQMC Circular Letter No. 242 of 6 May 1938.)

a. Officer's Cap and Hat Ornaments (par. 1 above).

CAP AND HAT ORNAMENTS

Line 3. Strike out "the garrison caps, ". • • In the same paragr aph, line 4, between "the" and "ornament" , insert "lef t service collar".

b. Officer's CoUar, Shoulder Knot and Epaulet Ornaments

(par. 4 above). COLLAR ORNAMENTS

At end of paragraph add: "Left service collar ornament shall be worn on the lef t front side of the garrison cap, with center of ornament two inches from the front."

c. Enlisted Cap, Hat and Collar Ornaments (par. 7 above).

CAP AND HAT ORNAMENTS

•• •line 3. Strike out "the garrison caps,"; and in same paragraph, in lines 4 and 5, strike out "bronze cap and hat ornaments" and insert "left service collar ornament".

10. Changes to Sections on Officer's Cap and Hat Ornaments; Officer's Full-Dress, Dress and White Cap Ornament; Officer's Summer and Winter Service Caps, Garrison Caps and Field Hat Ornament; Enlisted Cap, Hat and Collar Ornaments; Marine Band Cap, Hat, Collar and Shoulder Knot Ornaments.

(Extracts from HQMC Circular Letter No. 586 of 8 June 1942).

a. Officer's Cap and Hat Ornaments (par. 1 above).

CAP AND HAT ORNAMENTS

••.Strike out "AND HAT" (from caption). •••Line 2. Strike out "and hat". In line 3, strike out "and the field hat".

b. Officer's Full-Dress, Dress and White Cap Ornament (par. 2 above).

Ornament, cap, dress. - • • • line 1. Strike out "and hat".

c. Officer's Summer and Winter Service Caps, Garrison Caps and Field Hat Ornament (par. 3 above).

Ornament, cap and hat, service. - • • • line 1. In caption, strike out "and hat". In lines 20 and 21, strike out "and field hat".

d. Enlisted Cap, Hat and Collar Ornaments (par. 7 above).

CAP AND HAT ORNAMENTS

•••Strike out "AND HAT" (from caption). •••lines 2 and 3. Strike out "and hat". In lines 3 and 4, strike out "and the field hat".

e. Marine Band Cap, Hat, Collar and Shoulder Knot Ornaments (par. 8 above).

CAP AND HAT ORNAMENTS

• • • Strike out "AND HAT" (from caption.)

## APPENDIX K

## THE 1943 WOMEN'S RESERVE UNIFORM ORNAMENTS·

(Extracts from HQMC Letter of Instruction No. 523 ot 27 August 1943 with enclosure "Uniform Regulations, U. S. Marine Corps Women's Reserve, 1943".)

1. Officer's and Enlisted Ornaments.

   a. Cap and collar , officers, dress, shall be worn by officers with summer dress and summer service uniforms.

   b. Cap and collar , officers, service, shall be worn by officers with winter dress and winter service uniforms.

   c. Cap and collar , enlisted, dress shall be worn by enlisted personnel with summer dress, summer service B and summer service C uniforms.

   ct. Cap and collar , enlisted, service, shall be worn by enlisted personnel with winter service and summer service A uniforms.

2. Officer's Cap.( Hat and Collar Ornaments.

### CAP AND HAT ORNAMENTS

The service cap ornament shall be worn on the winter service cap, and dress cap ornament on summer dress and summer service caps.

### COLLAR ORNAMENTS

a. Service collar ornaments shall be worn on the winter service coat centered on the collar between the crease roll and outer edge with center of hemisphere 1 inch above the horizontal edge of lapel notch, the vertical axis of ornament being para llel with crease roll and eagle facing inside.

b. Dress ornaments shall be worn on the summer dress and summer service coats, centered on the collar at a point 3 3/4 inches from the point of the collar measuring along the outside edge and 1 3/4 inches from the outside edge measuring toward the notch, the vertical axis of ornament being parallel ·with crease roll and eagle facing inside.

3. Enlisted Cap, Hat and Collar Ornaments.

### CAP AND HAT ORNAMENTS

The dress cap ornament shall be worn on the summer cap; and the service cap ornament on summer service hat and winter service cap.

### COLLAR ORNAMENTS

a. Service collar ornaments shall be worn on winter service coat, centered on the collar between the crease roll and outer edge with center of hemisphere 1 inch above the horizontal edge of lapel notch, the vertical axis of ornament at right angles to the horizontal edge of notch, with eagle facing inside.

b. Dress collar ornaments shall be worn on summer dress coat, summer service B and summer service C coats, and service collar ornaments on summer service A coat, centered on the collar at a point 3 1/4 inches from the point of the collar measuring along the outside edge and 1 3/4 inches from the outside edge measuring toward the notch, the vertical . axis of ornament at right angles to the horizontal edge of notch , with eagle facing inside.

4. Change to Section on Officer's Cap, Hat and Collar Ornaments (par. 2 above). (

Extract from HQMC Letter of Instruction No. 670 of 25 February 1944.)

**COLLAR ORNAMENTS**

•• • Add "c. Dress collar ornaments shall be worn on the summer dress A coat on the center line of each shoulder strap, 3/4 inch from the arm- hole seam with head of eagle facing front."

5. Change to Sections on Officer and Enlisted Ornaments; Officer's Cap, Hat and Collar Ornaments; Enlisted Cap, Hat and Collar,- Ornaments.

(Extracts from HQMC Letter of Instruction No. 777 of 9 June 1944).

a. Officer and Enlisted Ornaments (para. 1 above).

• • •a. In second line, after "summer service", insert "B"•

• • •b. In second line, after "with", insert "summer service A,".

b. Officer's Cap, Hat and Collar Ornaments (par. 2 above).

CAP AND HAT ORNAMENTS

. • .In heading strike out "AND HAT". Strike out entire paragr aph, and insert. "The service cap ornament shall be worn on the winter service cap and the summer dress cap with summer service A uniforms. The dress cap ornament shall be worn on the summer dress cap with summer dress, summer dress A and summer service B uniforms. On the garrison cap the lef t service collar ornament shall be worn on the lef t side, in the place provided therefor."

COLLAR ORNAMENTS

a. In second line, after "coat", insert "and summer service Acoat".

b. In second line, after "service", insert "B".

c. Strike out entire par agraph and insert "c. Dress collar ornaments shall be worn on the summer dress A coat on the center line of each shoulder strap, center of hemisphere 3/4 inch from the armhole seam with head of eagle facing front",

c, Enlisted Cap, Hat and Collar Ornaments (par. 3 above),

CAP AND HAT ORNAMEN TS

•, . In heading, strike out "AND HA T". Strike out entire paragr aph, and insert c. "The service cap ornament shall be worn on the winter service cap and summer dress cap with summer service C uniform, The dress cap ornament shall be worn on the summer dress cap with summer dress and summer service B uniforms. On the garrison cap the lef t service collar ornament shall be worn on the lef t side, in the place provided therefor,"

COLLAR ORNAME N TS

, . ,b, Strike out entire paragraph and insert b, Dress collar ornaments shall be worn on summer dress coat and summer service B coat, and service collar ornaments on summer.. service A coat and summer service C coat, centered on the collar at a poin t 3 1/4 inches from point of the collar measuring along the outside edge and 1 3/4 inches from the outside edge measuring toward the notch, the vertical axis of ornament at right angles to the horizontal edge of notch, with eagle facing inside,

6, Change to Section on Officer and Enlisted Ornaments (par. 1 above),

(Extract from HQMC Letter of Instruction No. 803 of 18 July 1944,)

Ornaments, - c. In second line, after "summer dress", strike out comma and insert "and"; af ter ' ' summer service B", strike out " and summer service C", , ,d, In second line, af ter "winter service", strike out "and", and insert ","; af ter "summer service A", insert "and summer service C".

## APPF. NDIX L

## THE 1945 WOME N'S RESE RVE UNIFORM ORN AME N TS

(Extracts from Unifor m Regulations, U. S. Marine Corps Women's Reserve, 1945.)

1. Off icer and Enlisted Ornaments.

### ORNAME NTS

a.Cap and collar ornaments, off icers, dress, shall be worn by off icers with the summer dress and summer dress A unifor ms, and with the summer undress uniform when white pumps are worn.

b.Cap and collar ornaments, off icers, service, shall be worn by off icers with the winter dress , winter service, summer service, and summer par ade unifor ms, and with the summer undress uniform when brown pumps are worn.

c.Cap and collar ornaments, enlisted, gilt, shall be worn by enlisted persons wi th summer dress and summer undress uniforms,

d, Cap and collar ornaments, enlisted, bronze, shall be worn by enlisted persons with the winter dress, winter service, summer servtce , and summer par ade unifor ms,

2. Off icer's Cap and Collar Ornaments,

### CAP ORNAME NTS

a, The service cap ornament shall be worn on the winter service cap, and the summer dress cap with summer service and summer par ade unifor ms, and with the summer undress uniform when brown pumps are worn.

b, The dress cap ornament shall be worn on the summer dress cap wi th summer dress and summer dress A unifor ms, and with the summer undress uniform when white pumps are worn,

c. The lef t service collar ornament shall be worn on the garrison cap, on the lef t side, in the place provided therefor,

d, The cap ornament shall be worn with the eagle up,

### COLLAR ORNAME N TS

a, Service collar ornaments shall be worn on the winter service coat centered on the coll ar between the crease roll and outer edge with center of hemisphere 1 inch above the horizontal edge of lapel notch , the vertical axis of ornament being par allel with crease roll and eagle facing inside.

b. Collar ornaments shall be worn on the summer dress, summer service, and summer service L.S. coats centered on the collar at a point 3 1/4 inches from the point of the collar measuring along the outside edge and 1 3/4 inches from the outside edge measuring toward the notch, the vertical axis of the ornament being parallel with crease roll and eagle facing inside.

c. Dress collar ornaments shall be worn on the summer dress coat, and on the summer service L.S. coat, when the latter is worn with white pumps.

d. Service collar ornaments shall be worn on the summer service coat, and on the summer service L. S. coat, when prescribed as summer parade uniform; also when the summer service L. S, coat is worn with brown pumps,

e. Dress collar ornaments shall be worn on the summer dress A coat on the center line of each shoulder strap, center of hemisphere 3/4 inch from the armhole seam with head of eagle facing front,

3. Enlisted Cap and Collar Ornaments.

CAP ORNAMENTS

a. The bronze cap ornament shall be worn on the winter service cap, and on the summer dress cap with summer parade uniform.

b, The gilt cap ornament shall be worn on the summer dress cap with summer dress and summer undress uniforms,

c. The left bronze collar ornament shall be worn on the garrison cap, on the left side, in the place provided therefor.

d. The cap ornament shall be worn with the eagle up.

COLLAR ORNAMENTS

a. Bronze collar ornaments shall be worn on the winter service coat, centered on the collar between the crease roll and outer edge with center of hemisphere 1 inch aboe the horizontal edge of lapel notch, the vertical axis of ornament perpendicular, with eagle facing inside.

b. Collar ornaments shall be worn on the summer dress coat, summer service coat, and summer service L.S. coat, centered on the collar at a point 3 1/4 inches from point of the collar measuring along the outside edge and 1 3/4 inches from the outside edge measuring toward notch, the vertical axis of ornament perpendicular, with eagle facing inside,

c. Gilt collar ornaments shall be worn on the summer dress coat and on the summer service L. S. coat.

d. Bronze collar ornaments shall be worn on the summer service coat and on the summer service L.S. coat, when prescribed as summer parade uniform.

## APPENDIX M

## THE 1949 UNIFORM ORNAMENTS

(Extracts from Volume I, Marine Corps Manual, 1949.)

1. Male Officer's and Enlisted Men's Cap Ornaments.

**ORNAMENT, CAP**

a. The dress cap ornament shall be worn on the dress cap.

b. The bronze cap ornament shall be worn on the service caps and the fiber helmet.

c. The left service collar ornament shall be worn on the left front side of the service garrison caps, with the ornament centered vertically 2 inches from the front.

d. The design of the officer's cap ornament will be that of the United States Marine Corps emblem, omitting motto ribbon, consisting of Western Hemisphere approximately seven-eighths inch in diameter, intersected by a foul anchor and surmounted by an eagle; it will be provided with a screw post securely soldered to the hollow of globe projecting from the center, a and washer and milled nut for securing the ornament through eyelet to dress and service caps, and the fiber helmet. Enlisted personnel will wear the cap ornament as issued by the U. S. Marine Corps.

2. Male Officer's and Enlisted Men's Collar Ornaments.

ORNAMENT, COLLAR

a. Dress collar ornaments shall be worn on the blue dress and white coats, the evening dress and white mess jackets.

(1) On the blue coat, the white coat, and the evening dress jacket, the ornaments shall be placed vertically in the center of each side of the collar, the center of the hemisphere 1 1/2 inches from the neck opening, with the eagle facing to the front.

(2) On the white mess jacket, the ornaments shall be worn on the center line of the shoulder strap three-fourths inch from the armhole seam, with the eagle facing to the front and head toward the collar.

b. Gilt collar ornaments shall be worn by all enlisted men on the dress coat and by members of the Marine Band on the full-dress coat, except the full-dress coat worn by the Drum Major.

c. Service collar ornaments shall be worn on the service coats and service jackets. The service ornament shall be worn on the collar (in the eyelet provided therefor for enlisted men) centered between the crease roll and outer edge, with center of hemisphere 1 inch above the horizontal edge of lapel notch, the vertical axis of ornament at right angles to the bottom edge of the coat or jacket, with the eagle facing inboard.

d. Collar ornaments will not be worn on khaki shirts.

e. The design for the officer's collar ornaments will be that of the United States Marine Corps emblem, omitting motto ribbon, consisting of Western Hemisphere approximating eleven-sixteenths inch in diameter for the dress ornament, and approximately nine-sixteenths inch in diameter for the service ornament; it will be intersected by a foul anchor and surmounted by an eagle; it will be provided with a pin, screw, or clutch fastener for attaching the ornament to the dress and white coats, the evening dress and white dress jackets, the shoulder knots, the service coats, the jackets, and the garrison caps. Enlisted personnel will wear the collar ornaments as issued by the U. S. Marine Corps.

3. Women Officer's Cap Ornaments.

ORNAMENT, CAP, OFFICER

a. The bronze cap ornament shall be worn on the winter service cap, and on the summer dress cap with summer service and with the summer undress uniforms when brown pumps are worn.

b. The dress cap ornament shall be worn on the summer dress cap with summer dress and summer undress uniforms when white pumps are worn.

c. The left service collar ornament shall be worn on the left front side of the garrison cap in the place provided therefor.

4. Women Officer's Collar Ornaments.

ORNAMENTS, COLLAR, OFFICER

a. Service collar ornaments shall be worn on the winter service coat, centered on the collar between the crease roll and the outer edge, with center of hemisphere one inch above the horizontal edge of lapel notch, the vertical axis of ornament being parallel with crease roll and the eagle facing inboard.

b. Collar ornaments shall be worn on the summer dress "A", summer undress "A", and "B", and summer service coats, centered on the collar at a point 3 1/4 inches from the point of the collar, measuring along the outside edge, and 1 3/4 inches from the outside edge, measuring toward the notch, the vertical axis of the ornament being parallel with crease roll and the eagle facing inboard.

c. Dress collar ornaments shall be worn on the summer dress "A", summer undress "A", and summer undress "B" when worn with white pumps.

d.Service collar ornaments shall be worn on the summer service and summer undress "B" coats when worn with brown pumps,

e.Dress collar ornaments shall be worn on the summer dress "B" and summer undress "C" coats on the center line of each shoulder strap, center of hemisphere three-fourths inch from the armhole seam with the eagle facing to the front and head toward the collar.

5. Enlisted Women's Cap Ornaments.

OtlNAMENT, CAP, ENLISTED

a.The bronze cap ornament shall be worn on the winter service cap, and on the summer dress cap with summer service or summer undress "B" when worn with brown pumps.

b.The gilt cap ornament shall be worn on the summer dress cap with summer -dress and summer undress uniforms when white pumps are worn.

c.The left bronze collar ornament shall be worn on the left front side of the garrison cap in the place provided therefor.

6. Enlisted Women's Collar Ornaments.

ORNAMENTS, COLLAR, ENLISTED

a.Bronze collar ornaments shall be worn on the winter service coat, centered on the collar between the crease roll and outer edge with center of hemisphere 1 inch above the horizontal edge of lapel notch, the vertical axis of ornament perpendicular, with the eagle facing inboard,

b.Collar ornaments shall be worn on the summer dress, summer undress, and summer service uniforms, centered on the collar at a point 3 1/4 inches from point of the collar, measuring along the outside edge, and 1 3/4 inches from the outside edge, measuring toward notch, the vertical axis of orna- ment perpendicular, with the eagle facing inboard.

c.Gilt collar ornaments shall be worn on the summer dress coat, summer undress "A" and summer undress "B", when white pumps are worn.

d Bronze collar ornaments shall be worn on the summer service coat, and on summer undress "B", when brown pumps are worn.

7. Change to Section on Male Officer's and Enlisted Men's Collar Ornaments.

(Extract from Printed Change No. 4, January 1954 to Volume I, Marine Corps Manual, 1949.)

a.Collar ornaments will not be worn on khaki shirts, except when summer service "A" and "B" or winter service "C" is prescribed at which time enlisted men will wear the service collar ornaments centered on each side of the shirt collar midway between top edge of collar and point, ano .Jne (1) inch from edge of collar, with wings and eagle parallel to the top edge of the collar.

8. Change to Sections on Women Officer's and Enlisted Women's Cap and Collar Ornaments.

(Extracts from Printed Change No. 5, August 1954 to Volume I, Marine Corps Manual, 1949.)

a. Women Officer's Cap Ornaments.

ORNAMENT, CAP, OFFICER

(1)The bronze cap ornament shall be worn on the winter service cap and on the summer service cap with the uniforms prescribed, as applicable.

(2)The dress cap ornament shall be worn with the white dress and white undress uniform and on the blue cap.

(3)The left bronze collar ornament shall be worn on the left front side of the garrison, Dacron-Nylon, and utility caps in the place provided therefor.

b. Women Officer's Collar Ornaments.

ORNAMENTS, COLLAR, OFFICER

(1)Bronze collar ornaments shall be worn on the winter service coats and summer service uniforms in the eyelets provided therefor centered on the collar in the manner illustrated• . .

(2)Dress collar ornaments shall be worn on the white dress and white undress uniforms, and on the blue coat in the manner illustrated. • •

c. Enlisted Women's Cap Ornaments.

ORNAMENT, CAP, ENLISTED

(1)The bronze cap ornament shall be worn with the uniforms prescribed, as applicable.

(2)The gold cap ornament shall be worn on the summer service cap with white dress and white undress uniforms, an.ct on the blue cap.

(3)The left bronze collar ornar,ent shall be worn on the left front side of the garrison, Dacron-Nylon, and utility caps in the place provided therefor.

ct. Enlisted Women's Collar Ornaments.

ORNAMENTS, COLLAR, ENLISTED

(1)Bronze collar ornaments shall be worn on the winter service coats and summer service uniforms in the eyelets provided therefor centered on the collar in the manner illustrated• • •

(2)Gold collar ornaments shall be worn on the white dress and white un dress coat and on the blue coat in the manner illustrated• •• •

9. Change to section on Male Officer's and Enlisted Men's Collar Ornaments.

(Extract from Printed Change No. 6 , December , 1954 to Volume I, Marine Corps Manual , 1949.)

a.Service collar ornaments shall be worn on the service coats and ser- vice jackets. The service ornament shall be worn on the collar (in the eyelet provided therefor for enlisted men) centered between the crease roll and outer edge, with center of hemisphere 1 inch above the horizontal edge of lapel notch , the wing tips parallel to the bottom edge of the coat or jacket, with the eagle facing inboard.

b.Collar ornaments will not be worn on khaki shirts, except when summer service "A" and "B" is prescribed, at which time enlisted men will wear the service collar ornaments centered on each side of the shirt collar mid- way between top edge of collar and point , and one (1) inch from .edge of collar, with wings and eagle parallel to the top edge of the collar.

10. Change to section on Male Officer's and Enlisted Men's Collar Ornaments.

(Extract from Printed Change No. 9 , August , 1955 to Volume I, Marine Corps Manual , 1949.)

a.On the white mess jacket , the ornaments shall be worn on the collar , in the eyelets provided therefor , centered between the crease roll and outer edge, with center of hemisphere 1 inch above the horizontal edge of lapel notch, the wing tips parallel to the bottom edge of the jacket , with the eagle facing inboard.

11. Change to all sections on Male and Women Officer's and Male and Women Enlisted Cap and Collar Ornaments.

(Extracts from Printed Change No. 14 , June, 1956 to Volume I, Marine Corps Manual , 1949.)

a. Male· Officer's and Enlisted Men's Cap Ornaments.

INSIGNIA, BRANCH OF SERVICE

(1)The dress cap insignia, branch of service shall be worn on the dress cap.

(2)The bronze cap insignia, branch of service shall be wo. ı on the service caps and the fiber helmet.

(3)The lef t service collar insignia, branch of service shall be worn on the left front side of the service garrison caps, with the insignia, branch of service centered vertically 2 inches from the front.

(4)The design of the officer's insignia, branch of service will be that of the United States Marine Corps emblem, omitting motto ribbon , consisting of Western Hemisphere approximately seven-eighths inch in diameter , intersected by a foul anchor and surmounted by· an eagle; it will be provided with a screw post securely soldered to the hollow of globe projecting from the center , and a washer and milled nut for securing the insignia, branch .of service through eyelet to dress and service caps, and the fiber helmet. Enlisted personnel will wear the cap insignia, branch of service as issued by the U. S. Marine Corps.

b. Male Officer's and Enlisted Men's Collar

Ornaments. INSIGNIA, BRANCH OF SERVICE,

COLLAR

(1)Dress collar insignia, branch of service shall be worn on the blue dress and white coats, the evening dress and white mess jackets.

(a)On the blue coat, the white coat, and the evening dress jacket, the. insignia, branch of service shall be placed vertically in the center of each side of the collar , the center of the hemisphere 1 1/2 inches from the neck opening, with the eagle facing to the front.

(b)On the white mess jacket, the insignia, branch of service shall be worn on the collar , in the eyelets provided therefor, centered between the crease roll and outer edge, with center of hemisphere 1 inch above the horizontal edge of lapel notch, the wing tips parallel to the bottom edge of the jacket, with the eagle facing inboard.

(2)Gold collar insignia, br anch of service shall be worn by all enlisted men on the dress coat and by members of the Marine Band on the full-dress coat, except the full-dress coat worn by the Drum Major.

(3)Insignia, branch of service, collar bronze shall be worn on the green coats and green jackets. The service insignia shall be worn on the collar (in the eyelet provided therefor for enlisted men) centered between the crease roll and outer edge, with center of hemisphere 1 inch above the horizontal edge of lapel notch, the wing tips parallel to the bottom edge of the coat or jacket, with the eagle facing inboard.

(4)Collar inisgnia, branch of service will not be worn on khaki shirts, except when summer service "A" and "B" is prescribed, at which time enlisted men will wear the service collar insignia, branch of service centered on each side of the shirt collar midway between top edge of collar and point, and one (1) inch from edge of collar, with wings and eagle parallel to the top edge of the collar.

(5) The design for the officer's collar insignia, branch of service will be that of the United States Marine Corps emblem, omitting motto ribbon, consisting of Western Hemisphere approximately eleven-sixteenths inch in diameter for the dress insignia, branch of service, and approximately nine-sixteenths inch in diameter for the service insignia, branch of service; it will be intersected by a foul anchor and surmounted by an eagle; it will be provided with a pin, screw, or clutch fastener for attaching the insignia, branch of service to the dress and white coats, the evening dress and white dress jackets, the shoulder knots, the service coats, the jackets, and the garrison caps. Enlisted personnel will wear the collar insignia, branch of service issued by the U. S. Marine Corps.

c. Women Officer's Cap Ornaments.

INSIGNIA, BRANCH OF SERVICE FOR CAP, OFFICER

(1)The bronze cap insignia, branch of service shall be worn on the winter service cap and on the summer service cap with the uniforms prescribed as applicable.

(2)The dress cap insignia, branch of service shall be worn with the white dress and white undress uniform and on the blue cap.

(3)The left bronze collar insignia, branch of service shall be worn on the left front side of the garrison, Dacron-Nylon, and utility caps .in the place provided therefor.

d. Women Officer's Collar Ornaments.

**INSIGNIA, BRANCH** OF SERVICE, COLLAR, OFFICER

(1) Bronze collar insignia, branch of service shall .be worn on the winter service coats and summer service uniforms in the eyelets provided therefor centered on the collar in the manner illustrated.

(2) Dress collar insignia, branch of service shall be worn on the white dress and white undress uniforms, and on the blue coat in the manner illustrated• • •

e. Enlisted Women's Cap Ornaments.

INSIGNIA, BRANCH OF SERVICE FOR CAP, ENLISTED

(1)The bronze cap insignia, branch of service shall be worn with the uniforms prescribed, as applicable.

(2)The gold cap insignia, branch of service shall be worn on the summer service cap with white dress and white undress uniforms, and on the blue cap.

(3)The left bronze collar insignia, branch of service shall be worn on the left front side of the garrison, Dacron-Nylon, and utility caps in the place provided therefor.

f. Enlisted Women's Collar Ornaments.

INSIGNIA, BRANCH OF SERVICE, COLLAR, ENLISTED

(1)Bronze collar insignia, branch of service shall be worn on the winter service coats and summer service uniforms in the eyelets provided therefor centered on the collar in the manner illustrated.

(2)Gold collar insignia, branch of service shall be worn on the white dress and white undress coat and on the blue coat in the manner illus- trated• • •

12. Change to all sections on Male Officer's and Enlisted Men's Cap and Collar Ornaments.

(Extracts from Printed Change No. 44, February 1961 to Volume I, Marine Corps Manual, 1949.)

INSIGNIA, BRANCH OF SERVICE

a.Insignia, branch of service includes a cap device for wear with the frame cap, and a right and left device for wear on the collar of coats, jackets, and shirts as a pair. In addition a single left collar device will be worn on the left front side of the service garrison caps. Gold colored insignia will be worn on dress uniforms, and is referred to herein as dress insignia. Bronze insignia will be worn on all service uniforms, and is referred to herein as service insignia. Dress insignia will be maintained in a serviceable condition and will be kept highly polished. Service insignia will be maintained in a serviceable condition and will not be polished. Should continued use cause loss of the desired bronze finish, insignia will either be replaced, or refinished with USMC approved liquid, bronze, protective coating available through the Marine Corps exchange.

b.The design of the enlisted insignia, branch of service will be as issued or sold through the Marine Corps supply system. The design of the officers insignia, branch of service, cap both dress and service will be that of the USMC emblem without motto ribbon. It consists of a view of he globe showing Western Hemisphere approximately 7/8 inch in diameter, inter- sected by a foul anchor and surmounted by an eagle. It will be provided with a screwpost securely soldered to the hollow of the globe projecting from the center, and fitted with a washer and milled nut. The design of the officers dress collar insignia will be identical to the cap insignia except that it will be approximately 11/16 inch in diameter and will be provided with a pin, screw or clutch fastener. The officers insignia, collar service will be identical to the dress collar insignia except that it will be 9/16 inch in diameter.

c.The dress cap insignia will be worn on the cap frame when the white cover is worn with dress uniforms. It will also be worn on military police helmets. The service cap insignia will be worn on the cap frame when the green and \{haki covers are worn with the service uniforms. It will also be wor n on the fiber sun helmet.

d.Officers dress collar insignia will be worn on the blue and white coats and the evening dress and mess dress jackets. On the evening dress jacket the insignia will be placed vertically in the center of each side of the collar , the center of the hemisphere 1 3/4 inches from the neck opening with the eagle facing inboard. On the blue and white coats the insignia will be placed as above except the center of the hemisphere will be 1 7/8 inches from the neck opening. On the white mess jacket, the dress collar insignia will be worn on the collar in th,' eyelets provided therefor, centered between the crease roll and outer edge, with the center of the hemisphere 1 inch above the horizontal edge of the lapel notch, wing tips par allel, and the eagle f acing inboard. Enlisted dress collar insignia will be worn by enlisted personnel on the dress coat, and on the khaki shirt when the blue uniform is worn with- out coat. Members of the Marine Corps Band, except the Drum M ajor, will wear dress collar insignia on the full dress coat.

e. Officers insignia, branch of service, collar , service will be worn on the collar of the service coats and jackets, centered on the collar with the of the center hemisphere 1 inch above the horizontal edge of the lapel the wing notch, tips parallel to the bottom edge of the coat or jacket with the eagles facing inboard. Enlisted insignia branch of service collar ser- vice will be worn on service coats and jackets as above. In addition the collar service insignia will be worn on khaki shirts when worn as the outer garment as part of a service uniform. When the Winter Service "C" uniform is worn by enlisted personnel, insignia branch of service, collar , service will be worn on the collar of the khaki shirt in the manner p·re- scribed below. Commanding officers may authorize the wearing of collar insignia on both the shirt collar and the collar of the coat or jacket in order that personnel who contemplate removal of the coat or jacket during duty hours will be in proper uniform. When worn on the long sleeve khaki shirts the insignia will be centered on each side of the shirt collar midway between top and bottom edge of collar and one inch from the front edge, with the wing tips parallel to the top edge of the collar and the eagles facing inboard. When worn on the short sleeve khaki shirt, the insignia will be centered on a line bisecting the angle of the peak of the collar , with the center of the globe approximately 1 1/2 inches from the front and bottom edges of the collar, eagles facing inboard with the wing tips of the eagle parallel to the top edges of the shirt pocket when the shirt collar is open.

# APPENDIX N

## THE 1962 UNIFORM ORNAMENTS

(Extracts from Marine Corps Order P1020.34A of 2 October 1962 - Marine Corps Uniform Regulations.)

1. Male Officer's Cap and Collar Ornaments

### INSIGNIA, BRANCH OF SERVICE

a. Design

(1)The design of the officers dress cap insignia and the service cap insignia will be that of the Marine Corps emblem without motto ribbon. It consists of a view of the globe showing the Western Hemisphere approximately seven-eighths of an inch in diameter intersected by a fouled anchor and surmounted by an eagle. It will be provided with a screw- post securely soldered to the hollow of the globe projecting from the center , and fitted with a washer and milled nut.

(2)The design of the officers dress collar insignia will be identical to the cap insignia except that it will be approximately eleven-sixteenths of an inch in diameter and will be provided with a pin screw or clutch fastener.

(3)The design of the officers service collar insignia will be identical to the dress collar insignia except that it will be nine-sixteenths of an inch in diameter.

b. Cap Insignia

(1)The dress cap insignia shall be worn with the cap frame, centered in the eyelet provided therefor, with all dress or undress uniforms.

(2)The service cap insignia shall be worn when the cap frame is worn with service uniforms. It shall also be worn on the fiber sun helmet and the campaign hat.

(3)The left service collar insignia shall be worn on the left front side of the service garrison caps, with the insignia centered vertically 2 inches from the front.

c. Collar Insignia

(1)The dress collar insignia shall be worn on the blue and white coats and the evening and mess dress jackets.

(a)) On the blue and white coats the insignia will be placed vertically in the center of each side of the collar. The insignia will be placed in the eyelet provided therefor with the center of the hemisphere 1·7/8 inches from the neck opening with the eagles facing inboard.

(b)On the evening dress jacket, the insignia shall be placed as indicated above , except the center of the hemisphere will be 13/4 inches from the neck opening.

(c)On the white mess jacket, the dress collar insignia will be worn on the collar in the eyelet provided therefor. The insignia will be centered between the crease roll and edge, eagle facing inboard, the center of the hemisphere 1 inch above the horizontal edge of the lapel notch with the wing span of the eagles par allel thereto.

(2) Insignia, br anch of service, collar , bronze shall be worn on the collar of the service coats, centered on the collar with the center of the hemisphere 1 inch above the horizontal edge of the lapel notch, the wing tips par allel to the bottom edge of the coat and the eagles facing inboard.

2. Women Officer's Cap and Collar Ornaments.

INSIGNIA, BRANCH OF SERVICE

a.The design of the insignia, branch of service will be in accordance with the description set forth in par agraph 5105. (InRignia, Branch of Service for Male Officers.)

b. Cap Insignia.

(1)The dress cap insignia shall be worn with the dress caps, centered in the eyelet provided therefor.

(2)The service cap insignia shall be worn with the winter service cap and the green dress cap when it is worn with the summer service uniform, centered in the eyelet provided therefor.

(3)The lef t bronze collar iqsignia, branch of service shall be worn on the lef t front side of the summer service garrison and utility caps in the eyelet provided therefor.

c. Collar Insignia.

(1)The dress collar insignia shall be worn with blue and white coats and the evning dress jacket in the eyelets provided therefor , centered on the collar with the center of the hemisphere 1 inch above the hori- zontal edge of the lapel notch, the wing tips parallel to the bottom edge of the coat or jacket, with the eagle facing inboard.

(2)The insignia, branch of service, collar , bronze shall be worn on the winter service coat, summer service dress, summer service coat and in the same manner as prescribed above..

3. Enlisted Men's Cap and Collar Ornaments.

INSIGNIA, BRANCH OF SERVICE

a. The insignia, br anch of service; cap and collar , shall be the standard as issued or sold through the Marine Corps Supply System. In all cases the **insignia** shall be worn with the eagle facing inboard.

b, The dress cap insignia shall be worn on the frame, service cap when the white cap cover is worn. It will also be worn on military police helmets.

c.The service cap insignia shall be worn on the fiber helmet, campaign hat, and on the frame, service cap when the service cap cover is worn.

d.The dress collar insignia shall be worn on the dress blue coat in the eyelets provided thereon.

e.The service collar insignia shall be worn in the eyelets provided on the service coat or service jacket with the wing tips parallel to the bottom edge of the coat or jacket.

f.The left service collar insignia shall be worn on the lef t front side of the garrison cap, centered vertically 2 inches from the front.

4. Enlisted Women's Cap and Collar Ornaments.

INSIGNIA, BRANCH OF SERVICE

a.The insignia, branch of service, cap and collar , shall be the standard as issued or sold through the Marine Corps Supply System. In all cases, the insignia shall be worn with the eagle facing inboard with the wing tips parallel to the bottom edge of the coat or dress.

b.The dress cap insig;nia shall be worn on the blue dress cap in the eye- let provided.

c.The service cap insignia shall be worn on the winter service cap in the eyelet provided.

d.The dress collar insignia shall be worn on the dress blue coat in the eyelets provided.

e.The service collar insignia shall be worn on the summer service and winter service coats, and the summer service dress in the eyelets provided.

f.The lef t service collar insignia shall be worn on the left front side of the garrison caps. On the utility cap, it shall be centered vertically 2 inches from the front. On the summer service cap, it will be worn in the eyelet provided.

5. Change to the section on Enlisted Men's Cap and Collar Ornaments. (Para 3-b above.)

(Extract from MCO P1020.34A Change 2 of 7 February 1964.)

a. The dress cap insignia shall be worn on the frame, service cap when the white cap cover is worn.

# APPENDIX 0

## THE 1966 UNIFORM ORNAMENTS

(Extracts from Marine Corps Order P1020.34B of22 April 1966 - Marine Corps Uniform Regulations.)

1. Male Officer's Cap and Collar Ornaments.

### INSIGNIA, BRANCH OF SERVICE

a. Design

(1)The design of the officers' dress cap insignia and the service cap insignia will be that of the Marine Corps Emblem without motto ribbon. It consists of a veiw of the globe showing the Western Hemisphere approximately seven-eighths of an inch in diameter intersected by a fouled anchor and surmounted by an eagle. It will be provided with a screw- post securely soldered to the rear of the insignia at a point in the approximate center of the globe projecting from the center , and fitted with a milled nut.

(2)The design of the officers' dress collar insigni a will be identical to the cap insignia except that it will be approximately eleven-six- teenths of an inch in diameter and will be provided with a pin screw or clutch fastener.

(3)The design of the officers' service collar insignia will be identical to the dress collar insignia except that it will be nine-sixteenths of an inch in diameter.

b. Cap Insignia.

(1)The dress cap insignia shall be worn with the dress cap, centered vertically in the eyelet provided therefor , with all dress uniforms.

(2)The service cap insignia shall be worn when the cap frame is worn with service uniforms. It shall also be worn on the fiber sun helmet and the campaign hat.

(3)The lef t service collar insignia shall be worn on the left front side of the garrison caps, with the insignia centered vertically in the eyelet provided.

c. Collar Insignia

(1)The dress collar insignia shall be worn on the blue and white coats and the evening and mess dress jackets.

(a)) On the blue and white coats, the insignia will be placed vertically in the center of each side of the collar. The insigni a will be placed in the eyelet provided therefor with the center of the hemisphere 1 7/8 inches from the neck opening with the eagles facing inboard.

(b)On the evening dress jacket, the insignia shall be placed as indicated above, except the center of the hemisphere will be 1 3/4 inches from the neck opening.

(c)On the white mess jacket, the dress collar insignia will be worn on the collar in the eyelet provided therefor. The insignia will be centered between the crease roll and edge, eagle facing inboard, the center of the hemisphere 1 inch above the horizontal edge of the lapel notch with the wing span of the eagles parallel thereto.

(2) Service collar insignia shall be worn on the collar of the service coats, centered on the collar with the center of the hemisphere 1 inch above the horizontal edge of the lapel notch , the wing tips parallel to the bottom edge of the coat and the eagles facing inboard.

2. Women Officers' Cap and Collar Ornaments.

INSIGNIA, BRANCH OF SERVIC E

a.The design of the insignia, branch of service, will be in accordance with the description set forth in paragraph 5105. (Insignia Branch of Service for Male Officers.)

b. Cap Insignia

(1)The dress cap insignia shall be worn with the dress caps, centered in the eyelet provided therefor.

(2} The service cap insignia shall be worn with the winter service cap and the green dress cap when it is worn with the summer service uni- form, centered in the eyelet provided therefor.

(3) The lef t bronze collar insignia, branch of service, shall be worn on the lef t front side of the summer service garrison and utility caps in the eyelet provided therefor.

c. Collar Insignia

(1)The dress collar insignia shall be worn with blue and white coats and the evening/mess dress jackets in the eyelets provided therefor, centered on the collar with the center of the hemisphere 1 i,nch above the horizontal edge of the lapel notch , the wing tips parallel to the bottom edge of the coat or jacket, with the eagle facing inboard.

(2)The insignia, br anch of service, collar , bronze, shall be worn on the winter service coat, summer service dress, summer service coat and in the same manner as prescribed above.

3. Enlisted Men's Cap and Collar Ornaments.

INSIGNIA, BRANCH OF SERVICE

a.The insignia, branch of service, cap and collar , shall be the stanua:rd as issued or sold through the Marine Corps Supply System. In all cases, the insignia shall be worn with the eagle facing inboard.

b. The dress cap insignia shall be worn on the dress cap.

c.The service cap insignia shall be worn on the fiber helmet, campaign hat, and on the service cap.

d.The dress collar insignia shall be worn on the collar of the blue dress coat in the eyelets provided thereon.

e.The service collar insignia shall be worn in the eyelets provided on the service coat or service jacket with the wings parallel to the bottom edge of the coat or jacket.

f.The lef t service collar insignia shall be worn on the lef t front side of the garrison cap, centered vertically in the eyelet provided.

4. Enlisted Women's Cap and Collar Ornaments.

INSIGNIA, BRANCH OF SERVICE

a.The insignia, branch of service, cap and collar, shall be the standard as issued or sold through the Marine Corps Supply System. In all cases, the insignia shall be worn with the eagle facing inboard with the wing tips parallel to the bottom edge of the coat or dress.

b.The dress cap insignia shall be worn on the blue dress cap in the eyelet provided.

c.The service cap insignia shall be worn on the winter service cap in the eyelet provided.

d.The dress collar insignia shall be worn on the blue dress coat in the eyelets provided.

e.The service collar insignia shall be worn on the summer service and win ter service coats, and the summer service dress in the eyelets provided.

f.The lef t service collar insignia shall be worn on the lef t front side of the garrison caps. On the utility cap, it shall be centered vertically 2 inches from the front. On the summer service cap, it will be worn in the eyelet provided.

5. Change to the section on Enlisted Women's Cap and Collar Ornaments. (Para. 4-f above.)

(xtract from MCO P1020.34B Change 2 of 9 April 1968.)

a. The lef t service collar insignia shall be worn on the lef t front side of the garrison caps in the eyelet provided.

Chased - Ornamented Þy engraving;, embossing, or cutting.

Fretted - J\n ornamental pattern of small straight lines.

1"oul Anchor - The condition of an anchor when the chain , cable , or rope, has taken a tu rn around the flukes, shank, or stock.

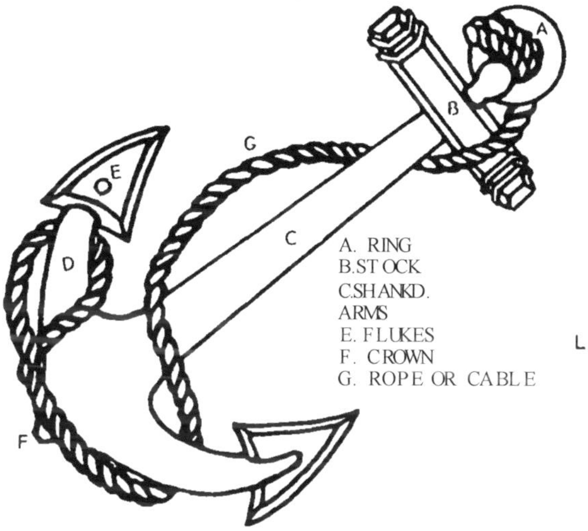


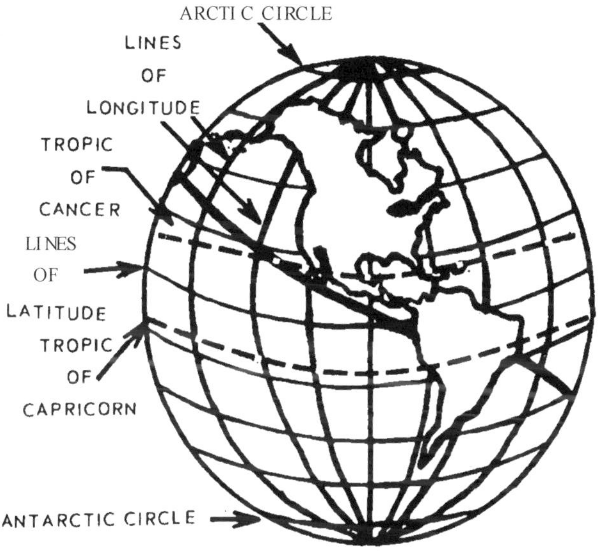


## TYPES OF FASTENERS

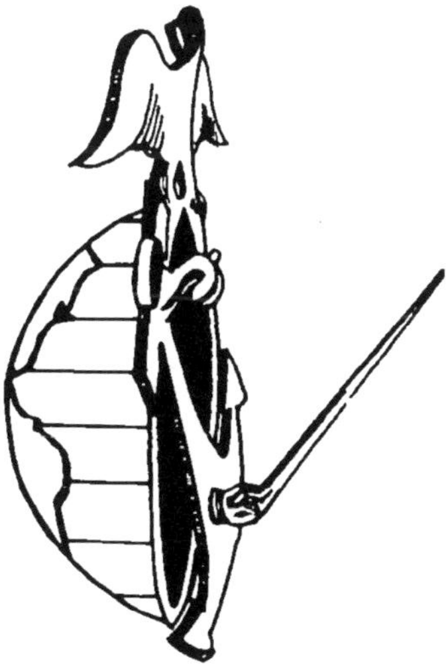

PIN FASTENER

SCREW FASTENER

CLUTCH FASTENER



☆ U. S. GOVERNMENT PRINTING OFFICE : 1977 246-686/1078

The device reproduced on. the back cover is the oldest military insignia in continuous use in the United States. It first appeared, as shown here, c Marine Corps buttons adopted in 1804. With the stars changeo to five points, this device has continued on Marine buttons to the present day.

Made in United States
Orlando, FL
04 November 2026

90204186R00101